Fixing Health Budgets: Experience from Europe and North America

Fixing Health Budgets: Experience from Europe and North America

Edited by

Friedrich Wilhelm Schwartz
Howard Glennerster
Richard B. Saltman

Technical editor

Reinhard Busse

JOHN WILEY & SONS
Chichester • New York • Brisbane • Toronto • Singapore

Baffins Lane, Chichester,
West Sussex PO19 1UD, England

National 01243 779777
International (+44) 1243 779777
e-mail (for orders and customer service enquiries): cs-books@wiley.co.uk
Visit our Home Page on http://www.wiley.co.uk
or http://www.wiley.com

Other Wiley Editorial Offices

John Wiley & Sons, Inc., 605 Third Avenue,
New York, NY 10158-0012, USA

Jacaranda Wiley Ltd, 33 Park Road, Milton,
Queensland 4064, Australia

John Wiley & Sons (Canada) Ltd, 22 Worcester Road,
Rexdale, Ontario M9W 1L1, Canada

John Wiley & Sons (Asia) Pte Ltd, 2 Clementi Loop #02-01,
Jin Xing Distripark, Singapore 0512

Library of Congress Cataloging-in-Publication Data

Fixing health budgets : experience from Europe and North America / edited by Friedrich Wilhelm Schwartz, Howard Glennerster, Richard Saltman. Technical editor Reinhard Busse.
p. cm.
Includes bibliographical references and index.
ISBN 0-471-96497-2 (cased : alk. paper)
1. Medical economics—North America. 2. Medical economics––Europe. 3. Medical policy—North America. 4. Medical policy––Europe. I. Schwartz, Friedrich Wilhelm. II. Glennerster, Howard. III. Saltman, Richard B.
RA410.F58 1996 94-52087
338.4′33621′094—dc20 CIP

British Library Cataloguing in Publication Data

A catalogue record for this book is available from the British Library

ISBN 0-471-96497-2

Typeset in 10/12pt Times from the editors' disks by Sally Vince, Frome, Somerset, UK
Printed and bound in Great Britain by Bookcraft, Midsomer Norton, Somerset, UK
This book is printed on acid-free paper responsibly manufactured from sustainable forestation, for which at least two trees are planted for each one used for paper production.

Contents

Contributors

Reinhard Busse is a Doctor of Medicine and Master of Public Health in the Department of Epidemiology and Social Medicine at the Medizinische Hochschule Hannover, D-30623 Hannover, Germany.

Johan Calltorp is Professor of Health Services Management at the Nordic School of Public Health, PO Box 12133, S-40242 Göteborg, Sweden.

Theodore G. Ganiats is Associate Professor and Head of the Division of Family Medicine in the Department of Family and Preventive Medicine, School of Medicine, University of California at San Diego, 9500 Gilman Drive, La Jolla, California 92093, USA.

Howard Glennerster is Professor of Social Policy and Chairman of the Suntory and Toyota International Centres for Economics and Related Disciplines at The London School of Economics and Political Science, University of London, Houghton Street, London WC2A 2AE, UK.

Chris Ham is Professor of Health Policy and Management and Director of the Health Services Management Centre in the School of Public Policy, University of Birmingham, 40 Edgbaston Park Road, Birmingham B15 2KT, UK.

Chris Howorth is a student of economics and social policy at Royal Holloway, University of London, UK, and the Department of Epidemiology and Social Medicine, Medizinische Hochschule Hannover, Germany.

Robert M. Kaplan is Professor and Head of the Division of Health Care Services in the Department of Family and Preventive Medicine, School of Medicine, University of California at San Diego, 9500 Gilman Drive, La Jolla, California 92093, USA.

Hans A.M. Maarse is Professor of Health Care Policy Analysis in the Faculty of Health Sciences at the University of Limburg, Postbox 616, NL 6200 MD, Maastricht, The Netherlands.

Alan Maynard is Secretary for The Nuffield Provincial Hospitals Trust for Research and Policy Studies in Health Services, 59 New Cavendish Street, London W1M 7RD, UK. Professor of Economics and Director of the Centre for Health Economics, University of York, until 1995.

Lawrence J. Nestman is Professor and Director of the School of Health Services Administration, Dalhousie University, 1234 Seymour Street, Halifax, Nova Scotia B3H 3M3, Canada.

Karl P. Pfeiffer is Professor the Institute for Bio-statistics and Documentation, University of Innsbruck, Anichstr. 35a, A-6020 Innsbruck, Austria, and Director of the Ludwig Boltzmann-Institute for Epidemiology and Health System Research in Innsbruck, Austria.

Sabine Richard is a Doctor of Economics at the Contracts Office of the Bundesverband der Betriebskrankenkassen (Federal Association of the Company-based Sickness Funds), Kronprinzenstr. 6, D-45128, Essen, Germany.

Richard B. Saltman is Professor in the Department of Health Policy and Management at the Emory School of Public Health, 1518 Clifton Road NE, Atlanta, Georgia 30322, USA.

Karl-Heinz Schönbach is Head of the Contracts Office of the Bundesverband der Betriebskrankenkassen, Kronprinzenstr. 6, D-45128, Essen, Germany.

Friedrich Wilhelm Schwartz is Professor of Epidemiology and Social Medicine, Director of the Department of Epidemiology and Social Medicine at the Medizinische Hochschule, D-30623 Hannover, Germany and Chairman of the North German Centre for Health Research.

Clive H. Smee is the Chief Economic Adviser at the UK government's Department of Health, Skipton House, 80 London Road, London SE1 6LW, UK.

Wynand P.M.M. van de Ven is Professor of Health Insurance in the Institute of Health Care Policy and Management, Erasmus University, PO Box 1738, NL 3000 DR, Rotterdam, The Netherlands.

Preface

Almost twenty years of voluntary budgets in Germany's ambulatory care sector and three years' experience with legally fixed budgets in several sectors of the German health care system provided the background for a two-day international seminar on "Health Care Management of Costs and Quality under Fixed Budgets" in Celle, Germany, in March 1995. Academics, health care managers and health care administrators from Austria, Canada, Germany, the Netherlands, Sweden, the United Kingdom and the United States discussed experience with fixed budgets in health care. General topics included intended and unintended effects on costs and quality, successes and failures, and other related issues.

First drafts of most chapters in this volume were discussed at that seminar, which was hosted by the Department of Epidemiology and Social Medicine and the North German Centre for Public Health Research, both at Medizinische Hochschule Hannover. The seminar was co-sponsored by the European Healthcare Management Association (EHMA). Seminar participants included Gunar Baugut, Jacob A. Bijkerk, Margot Fälker, Gunnar Griesewell, Reiner Leidl and Jörg-Volker Paschke.

ACKNOWLEDGEMENTS

The editors are grateful to the Volkswagen Foundation for a grant in support of the seminar, to the German Federal Ministry of Health for financial support of this publication, to Philip Berman from EHMA for his valuable support before, during and after the seminar, and to Chris Howorth who assisted in preparing the final manuscripts.

Introduction

The notion of a fixed health budget seems self-evident: a public or private payer sets a time-tied ceiling on expenditures. Such a definition, while correct, is none the less far from sufficient. In actual practice, the key issues revolve around not only the "what," but also the "who" and the "how" of establishing and maintaining periodic health care expenditure ceilings. As the chapters that follow demonstrate, the design and implementation of fixed budgets in health systems – not to mention their success – varies considerably between countries, within countries, and between health sub-sectors. Even if they succeed, what effect do they have? Do patients suffer? Do doctors tighten their belts and have to live a little more like ordinary mortals? Who gains? Who loses?

Governments typically adopt fixed health budgets for several related reasons. The first and most pressing, not surprisingly, tends to be domestic fiscal pressures. These manifest themselves in diverse ways. In insurance-based systems, governments worry about the burden on employers and workers of high contribution rates. In tax-based systems, governments worry about the political consequences of high taxes. Hence, governments try to limit the spending of their health care systems – as in Germany in 1989 with the imposition of fixed budgets for ambulatory physician visits, and in 1993 with the establishment of similar expenditure ceilings for ambulatory pharmaceuticals. In the UK, there has been permanent structural competition for public funds through an annual public budgeting process since the 1948 inception of the National Health Service. It has included a fixed budget, of a kind, ever since that time.

These domestic fiscal pressures, in turn, tend to reflect the full range of political issues that move national and/or regional governments – from concern about the long-term stability of health care revenues to short-term electoral cycles with their accompanying need to be seen to be "doing something." More recently, a central domestic political concern has been to improve national economic competitiveness in an era of both globalizing and regionalizing markets. This concern over international competitiveness has contributed in a number of developed countries to the current political desire to adopt a more stringent fiscal policy in the health sector generally.

THE MECHANISMS OF FIXING BUDGETS

Current experience demonstrates that fixed health budgets can be constructed along a variety of alternative dimensions. In different countries they are: established and administered by national governments (UK, Netherlands), by regional governments (Sweden, Canada), by non-governmental statutory agencies (Germany and France), or by private for-profit companies (USA). Budgets can be fixed for the entire health sector, for the entire publicly administered health sector (UK, Finland, Sweden), and/or for specific health sub-sectors such as hospitals (Netherlands), ambulatory physician visits (Germany) and pharmaceuticals (Germany), as well as for various combinations of these categories.

The mechanisms utilized to define and maintain fixed budgets are equally varied. Their design and adoption typically reflect the degree to which a fixed budget is intended to be a "hard" or a "soft" cap on overall operating expenditures – or, to phrase it a bit differently, whether the policy objective is to establish a fixed budget or a target budget. Similarly, the extent to which such budgets carry negative sanctions if they are breached, the willingness actually to apply those sanctions, and which agency or actor in fact monitors budgets and triggers the application of sanctions also vary with this more general goal.

Fixed budgets typically utilize numerous forms of expenditure control over provider behavior. Among the techniques used are capitation, case-based reimbursement instruments such as Diagnostic-Based Groups (DRGs), negotiated fee schedules, and a time-tied collective total for a category of providers (e.g. *Länder*-based three-month pools for ambulatory physician payments in Germany). Similarly, a variety of different contracting mechanisms can be employed: block contracts (for a general category of services), per case (for each episode), volume-based (for an agreed number of cases) and price and volume (creating payment corridors contingent on volume). Further, all these types of contracts can be hard (adversarial) or soft (retrospectively adjusted) in nature (von Otter, 1991).

These mechanisms focus on provider operating behavior; others can be adopted to constrain capital development – reflecting the role of capital expenditures for buildings, capital equipment, and new services in driving up subsequent operating costs. Such mechanisms include restrictions on the amount and/or sources of new capital (e.g. only from pooled public funds in Sweden); restrictions on new capital outlays for buildings, renovations, and/or capital equipment over a certain threshold (e.g. Article 18 permits in the Netherlands, certificate-of-need programs in some US states); restrictions on the number of medical students and/or residents (e.g. *numerus clausus* in Germany; a government quota on medical student places in the UK); caps on the number of authorized provider-related civil servant posts (e.g. health personal in Finland prior to 1993); and restrictions on the number and/or utilization of beds in a particular provider institution (e.g. special permits for intensive care facilities).

A central characteristic of all these mechanisms is the desire to generate appropriate behavioral incentives for physicians and provider institutions to honor their fixed budget agreements. Beyond financial carrots and sticks, incentives that regulators seek to harness include peer respect and peer regulation as well as professional satisfaction for physicians, and peer review status and quality of service and reputation for hospitals.

In certain self-regulatory regimes like that of Germany, providers also may have strong incentives to ensure that existing frameworks remain viable so as to maintain their self-regulatory and thus formally independent status. This raises the general question about the role of intermediate institutions in the construction and operation of fixed budgeting mechanisms. Political commentators since Aristotle have spoken of the importance of having either political or social bodies that can protect the individual from the full weight and authority of the State. In federal countries such as Germany, a central factor in the design of health financing mechanisms generally has been the self-regulatory character of both sickness funds and providers, and the importance of buffering the direct influence of the State on both providers and patients.

THE PROCESS OF FIXING BUDGETS

Regardless of the specific structure adopted, each fixed budget arrangement will involve a continuing dynamic of design and implementation. This essentially political process has five relatively consistent elements: development, approval, monitoring, evaluation, and sanctions. These different stages are subject to various political cycles – most notably the short-term electoral needs of the party in power. They also reflect the varying lobbying relationships among key actors (in particular, between physicians and political leaders), and the impact these relationships may carry for each of the five stages of the policy process.

A critical factor in the process of establishing fixed budgeting in any national context is that of developing adequate information systems. Data about price, quality, financial performance, managerial efficiency, and the impact on equity and health gain all provide essential guideposts for an effective budgeting system, and also improve the transparency of the outcomes to both providers and patients.

A last but equally important aspect of the process of fixed budgeting concerns the issue of priority setting and rationing of services. How to allocate scarce resources has become a central part of many national debates about controlling health expenditures. In the Netherlands and Sweden, there have been national reports on the subject of priority setting (Ministry of Welfare, Health, and Culture, 1992; Health Care and Medical Policy Commission, 1993). Chris Ham suggests in his chapter in this volume (Chapter 2) that priority setting will become an increasingly important factor in the ability of governments to manage fixed budgets.

JUDGING FIXED BUDGETS

Perhaps the most controversial question about fixed budgets in the health sector is "Do they work?" Part of the controversy reflects different criteria that analysts bring to making that judgement.

It would appear that both the positive and negative aspects of fixed health budgets have these basic components: efficiency or cost, equity, and quality. In considering its positive impact, efficiency should be assessed in terms of aggregate cost containment, cost-effectiveness of clinical care, and dynamic efficiency within institutions. In terms of its negative impact, it is important to consider the extent of any cost-shifting on to other public, private, or household budgets, or whether higher transaction costs were generated. The positive elements of equity ought to be evaluated in terms of the allocative efficiency across the service delivery system and the potential for greater transparency to support leveling across the system. The negative elements ought to include consideration of cream-skimming by either third-party-payers or providers (both adverse selection and selective disenrollment), as well as indicators of a two-tier-system of service or quality. Regarding the third criterion, quality, positive effects might include reductions in unnecessary and non-cost-effective treatments, and pressure to revise organizational and professional systems of quality control. Negative effects would include opportunistic behavior by providers to reduce quality or the development of a broadly cost-driven rather than quality-driven health system.

THE CHAPTERS

The chapters in this book have been grouped in accordance with three levels of health system configuration: macro or system level; sectoral and regional level; and micro or institutional provider level. Each must be involved in the budget-fixing process to some degree but the emphasis varies between health care systems. In what follows, practitioners and academics outline how far their countries have moved to adopt fixed budgets and what difficulties they encountered.

REFERENCES

Health Care and Medical Policy Commission (1993). *No Easy Choices – Difficult Priorities in Health Care*. Stockholm. Ministry of Health and Social Affairs. Swedish Government Official Report, 1993:93.

Ministry of Welfare, Health, and Culture (1992). *Choices in Health Care*. Rijswijk.

von Otter, C. (1991). The application of market principles to health care. In: D. Hunter (Ed.), *Paradoxes of Competition for Health*. Nuffield Institute, University of Leeds, UK, pp 13–22.

Part 1

System Level Issues

Chapter 1

Thinking about Planned Markets and Fixed Budgets

Richard B. Saltman

THE NOTION OF PLANNED MARKETS

The idea of a planned market is a straightforward one (Saltman and von Otter, 1992). Specific market-style mechanisms can be extracted from a neoclassical economic model, stripped of their link to private ownership of capital, and injected into an existing publicly planned, bureaucratically administered health system. A variety of different market mechanisms can be utilized: consumer sovereignty (patient choice), negotiated contracts (managerial choice), open bidding, and so forth. The market incentives can be introduced on the financing (funding) side of the health system, on the production side, or in the design of the allocation mechanism (the provider payment system) that connects financing to provision (Saltman, 1994b).

The objective in introducing market mechanisms into bureaucratically administered health systems is to change the system's behaviour without changing its outcome. That is, the goal is a more efficient means to achieve the same, previously pursued end. This end, of course, continues to represent a publicly defined, publicly accountable purpose: equal access for all citizens to timely, appropriate, high-quality care.

The use of market mechanisms to achieve public sector objectives directly contradicts the assumptions of traditional economic theory about the proper role and structure of a market. In pure neoclassical economics, markets function best when composed of privately capitalized, privately managed enterprises that participate in the market explicitly and solely to maximize their private financial return (e.g. profit). Neoclassical economists such as Milton Friedman and Friederich von Hayek rail against any governmental attempt to rein-in or moderate market behaviour (Friedman, 1962). Present-day health economists tend to draw on this logical foundation to argue that publicly operated health systems are "monopolies" that necessarily are less efficient than a market-organized delivery system would be (Rehnberg, 1995).

Fixing Health Budgets: Experience from Europe and North America.
Edited by F. W. Schwartz, H. Glennerster and R. B. Saltman. Technical editor R. Busse.

A planned market, conversely, draws on a school of political economists that believes that markets exist within a social and political context which, by definition, constrains and channels their behaviour. Among the oldest proponents of this position is, of course, Adam Smith. His *The Theory of Moral Sentiments* (Smith, 1759) followed other Scottish philosophers such as Robert Hutchinson in describing appropriate private economic behaviour as reflecting the moral expectations of the broader community within which it took place.

Smith wrote in this regard that man:

> has a natural love for society, and desires that the union of mankind should be preserved for its own sake, and though he himself was to derive no benefit from it. The orderly and flourishing state of society is agreeable to him, and he takes delight in contemplating it. (Smith, 1759, p 88)

Several centuries later, Karl Polanyi contended that economic behaviour necessarily occurred within boundaries established by "society" – by the State – in order to ensure socially tolerable as well as economically positive outcomes (Polanyi, 1944). Polanyi argued powerfully that no society could allow "a market system controlled, regulated, and directed by markets alone" (p. 68):

> To allow the market mechanisms to be the sole director of the fate of human beings and their natural environment, indeed, even of the amount and use of purchasing power, would result in the demolition of society. ... Robbed of the protective covering of cultural institutions, human beings would perish from the effects of social exposure. ... Nature would be reduced to its elements, neighbourhoods and landscapes defiled, rivers polluted. ... Finally, the market administration of purchasing power would periodically liquidate business enterprise, for shortages and surfeits of money would prove as disastrous to business as floods and droughts in primitive society. (Polanyi, 1944, p. 73)

Consequently, in the nineteenth century as in earlier and later history:

> a network of measures and policies was integrated into powerful institutions designed to check the action of the market relative to labor, land, and money. ... Society protected itself against the perils inherent in a self-regulating market system – this was the one comprehensive feature in the history of the age. (Polanyi, 1944, p. 76)

In Polanyi's analysis, modern markets are necessarily political markets, structured according to the expectations of sovereign governments. From this perspective, it is impossible to conceive of a socially or economically acceptable arrangement that could rightly be termed a "free market", since sustainable markets inherently require active intervention and regulation by the State.

A pragmatic variant of Smith's and Polanyi's analyses can be found in the regulatory theory that underlies the behaviour of the modern liberal state. In this analysis, a market left to its own devices would soon be manipulated by its largest participants, rapidly losing its competitive characteristics as oligopolistic and monopolistic activity came to predominate. Similarly, financial markets could readily be debased and discredited by unscrupulous or opportunistic firms. Consequently, advanced capitalist societies find it necessary to preserve the basic competitive characteristics of a market by establishing explicit antitrust and other transparency requirements monitored by a monopolies board or a competition commissioner. Here, too, as with Smith and Polanyi (if in less morally and politically explicit language), the argument is that economic markets do not, indeed cannot, exist for their own sake nor within a political and social vacuum. Rather, man's economic activities can successfully be pursued only within a broader moral and/or social framework developed and introduced by a political authority – by the State.

Viewed from this perspective, planned markets follow along on a well-trod political economist path. The overall goal of planned markets – to harness private competitive incentives to achieve stated public sector objectives – is in the tradition of past and present notions that communities and governments – collective political bodies – must and should legislatively define the boundaries and objectives of market activity. Just as European governments have decided to allow markets that trade in bananas, for instance, but to ban markets that trade in human beings (at least in recent centuries) or in human organs, the argument for planned markets contends that public sector bodies ought to define what areas of a health system can appropriately be opened up for market-style behaviour, and what areas should remain closed to any type of competitive incentives.

Moving beyond broad economic theory, planned markets also can be assessed in terms of recent theories about management of large private sector firms (Saltman and von Otter, 1995). In this perspective, the entire health system can be understood as an integrated production structure, in effect paralleling the behaviour of a large vertically integrated corporation. The currently popular idea of "re-engineering" corporations (Hammer and Champy, 1993), for example, shares with the notion of planned markets such key objectives as greater operational flexibility, improved productivity and a more client-centred focus (although a planned market need not generate the permanent job insecurity that typically accompanies re-engineering in the private sector). Similarly, the concept of "strategic framing" (Hamel and Prahalad, 1994) closely resembles the core planned market approach in which political policy objectives are set by senior elected officials, while subsequent operational decisions about how to achieve those objectives are the responsibility of individual provider institution managers (Saltman and von Otter, 1992). These two management strategies can be viewed as reinforcing the planned market argument that publicly accountable health systems can better achieve their existing objectives through flatter administrative hierarchies combined with more managerially independent and entrepreneurial service delivery units.

The notion of a planned market, then, can be understood to draw on key concepts from both political economics and present-day management theory. In this sense, it represents the application of both macro/system and micro/institutional lines of academic logic to the conundrum of health system renewal.

THE STRUCTURE OF PLANNED MARKETS

There are a variety of different ways to structure a planned market for health services.[1] An initial typology developed distinctions between a mixed market as contrasted to a public competition approach to planned markets (Saltman and von Otter, 1989). A subsequent analysis underscored the role of transaction costs and emphasized the relative advantage in constructing a planned market on "soft", trust-based rather than "hard", adversarially based contracts between health care purchasers and providers (von Otter, 1991, 1994). Present planned market developments in Northern European health systems suggest that, in addition, a standard business school framework for analysing private sector business strategies might be useful. Drawing upon this framework, national policy makers have the option to construct a planned market in health care in which competition is based predominantly on price, on quality, or on market share. As in the private sector, each of these different competitive strategies has its advantages and disadvantages, and each has its champions and its critics.

Constructing a planned-health care market based on price competition has the advantages, for example, of stimulating dynamic efficiencies to lower prices and of encouraging innovation in service delivery. It is the type of planned market that is usually based on negotiated contracts between purchasers and providers. Economists typically favour a price-based market, arguing that it can introduce market discipline and, through competitive behaviour, reduce the price for delivered health services.

Yet a price-based planned market also can generate a variety of social and health-related problems. Among the most serious is encouraging providers to adversely select among potential patients. Another disadvantage of this price-based, negotiated contract model is that the patient typically has little if any choice of specialist or hospital at which secondary and tertiary care is provided – expressed in financial terms, the patient must follow the money. Further, the greater the contract-generated pressure on providers to find ways to reduce prices, the greater the temptation of providers to cut corners with regard to quality – to engage in what transaction economists refer to as opportunistic behaviour (Williamson, 1975). Eventually, price competition runs the danger of steering a market toward a standard but rather low level of health care – what might be termed the McDonaldization of the market. There also remains the potential for higher administrative and other transaction costs for the entire system. In particular, it can be very expensive to generate and update real prices in a complex service sector like health.

Price-based forms of planned markets often incur opposition from advocates for public health policies, as well as from decision-makers committed to equity of access and/or to patient choice (Diderichsen, 1992; Ham, 1992). These opponents worry that resource allocation decisions in the health sector will no longer adequately reflect societal and population-based objectives that they believe are either more important than or, in certain cases such as some types of prevention, could well provide additional reinforcement to, specifically financially oriented goals. Opponents of price-based planned markets also contend that a sound understanding of transaction economics indicates that price-based markets are unlikely to achieve the promised financial efficiencies (Saltman and von Otter, 1989).

Planned markets in which competition is based predominantly on quality have roughly the converse characteristics, and thus an opposite roster of proponents and critics. Predominantly quality-based planned markets involve establishing a fee schedule for services, usually by state or regional policy makers, that rely on only an approximation of real prices. The competitive element is brought in by patients, who select the physician or institution they prefer (for secondary and tertiary care, typically with the counsel of their primary care provider) at which to receive services. This approach, in theory, allows patients to accumulate clinical information on the quality of services that different providers offer, and also logistical information on waiting times, appointment procedures, and other scheduling issues at competing provider institutions. From a financial perspective, it means that the money follows the patient, and thus that providers are paid based on some relationship to the volume of patients they attract (e.g. volume corridors, DRGs, capitation, etc.).

The central advantage of a predominantly quality-based planned market is that universality of access to care is relatively easier to achieve, in that patients control decisions about which provider to seek out. Further, providers find that it is in their best economic interests to practise with sufficient quality of care to ensure a reasonable recommendation from their patients (Saltman, 1994a). Depending on the complexity of the fee schedule established, a quality-based planned market should have relatively low administrative and other transaction costs, reflecting the fact that a fixed-price market is less expensive to operate.

Public health advocates and solidarity-oriented policy makers tend to find this type of planned market arrangement more attractive than price-based approaches. They believe that it can ensure continued political steering of resource allocations so as to better protect population and needs-based programmes. Of course, some public health advocates dislike planned market models generally, preferring to combine the stability of bureaucratically administered systems with intensified managerial efforts (Ham, 1992).

The fixed-price nature of a predominantly quality-based planned market also carries disadvantages. As an artificially priced system of resource allocation, there is a danger that its administratively set prices may not reflect the real costs of providing care and that, consequently, distortions will develop in service availability and treatment patterns. Innovative services may be slower to start up, particularly if the pricing regimen is not flexible. An additional dilemma is that a patient-based market in health services requires an explicit mechanism to provide sufficient and appropriate information to overcome the naturally asymmetric character of medical knowledge (i.e. providers know more than patients) (Saltman, 1994a). All these disadvantages can lead to the conclusion that quality-based markets are not adequately responsive to short-term economic incentives and – not surprisingly – to disdain for this type of planned market from neoclassical economists.

The third design option for a planned market, based predominantly on market share, implies more the strategy of individual for-profit participants in either a price-based or a quality-based market than it does the intentional policy of a public planning body. In standard business school theory, organizations should seek to become one of the two or three largest volume providers of a product if they expect successfully to influence their financial future in a competitive environment (Porter, 1980). Typically, this requires the company either to underprice its competition, to provide a demonstrably superior product, or to otherwise out-market its competitors, ideally with sufficient ferocity to drive the weaker ones out of business. After acquiring a relatively large market share, a company will have the ability to establish "price leadership", that is, to increase its prices and its profit margins. Clearly, this approach shares some of the advantages of both price-based and quality-based markets. It also, however, carries both strategies' disadvantages as well.

PLANNED MARKETS AND FIXED BUDGETS

The notion of a planned market has an in-built advantage with regard to controlling aggregate expenditures, in that it is derived from the fully planned, command-and-control model of health system organization that was its direct predecessor. A key characteristic of the command-and-control model, of course, was that it was grounded in a fixed system-level budget. Typically, command-and-control health systems were financed predominantly from publicly accumulated revenues, whether collected at a single level of government (as in the UK and Italy) or by a mix of two different governmental sources (as in Sweden and Finland). The annual budgets for command-and-control health systems typically were set at the beginning of the budget year, based on the anticipated tax revenues that would be available. Global budgets for specific provider units in particular hospitals were then derived, based on the total sum that would be available.

Several concerns have been raised, however, about the operation of fixed macro-level budgets in command-and-control systems. In the UK, for example, there were no individual caps (e.g. cash-limits) on general practitioners or on outpatient pharmaceutical/prescriptions. While this had the advantage of encouraging treatment of as many patients as possible at the less intensive, less expensive primary care level of the system, the cost of these outpatient services was not entirely predictable on a prospective annual basis – a circumstance which unsettled the Treasury. A further, arguably less positive effect of traditional fixed budgets was the tendency of British hospitals to shut wards down in the last few weeks or months of the budget year, as a technique for staying under budget.

Viewed from this historical perspective, one valuable characteristic of a planned market approach is that it can retain a relatively rigid fixed budget at the macro system level at the same time that it can make budgeting at the individual institutional level more flexible (Saltman and von Otter, 1992). Within a prospectively set macro system ceiling, institutional-level revenues can be tied to competitive incentives so as to encourage greater efficiency in daily operating behaviour and higher levels of effectiveness in the outcomes achieved. It also is possible to tailor the use of different competitive mechanisms to the needs of different sub-sectors within the health system.

There are both price-based and quality-based notions about how planned markets might achieve these objectives. The price-based approach calls for the design of a framework for negotiated contracts, within which providers are encouraged to reduce their prices in order to secure a guaranteed volume of cases. The purchasing authority, in turn, as well as its senior management, are rewarded financially if they can obtain lower than average expenditure rates. In this approach to planning the market for health services, providers that compete by reducing their prices will presumably be rewarded by winning a greater volume of business, and hence, at the end of the year, a larger percentage of (fixed) total expenditures.

A quality-based approach focuses not on purchasers and price, but rather on individual patient preferences as the key allocator of health sector resources. An internal set of prices and/or payments for specific services is established by the relevant political authority, typically on an annual basis, with all providers paid more or less the same amount for providing the same services (systems sometimes have historical or geographical adjustments). Subsequently, a provider's revenues are determined by the number of patients who select that provider, multiplied by the established amount per patient and/or treatment category. In 1988 in Sweden, for example, Stockholm County introduced a two-priced capitation system for primary care, with a fixed annual fee for individuals under age 65, with twice the annual fee for individuals over 65 (Saltman and von Otter, 1992). Primary health centres, in turn, received a budget based upon the number of patients on their list.

A central element of these planned market models is that only system-level expenditures are capped. In both the price-based and the quality-based approach, a particular provider doesn't have a fixed prospective global budget. Rather, in both instances, the resources that a provider has available will depend upon their success in bringing in patients, either through successful contract negotiation or patient choice.

In effect, each of these models removes the cap (and the floor) on each provider's annual revenues, encouraging individual providers to act competitively to improve their overall income. Since the total available revenue in either model is fixed at the aggregate level (or, one might imagine, at the sub-sector level, e.g. for hospital inpatient services), this competitive market cannot expand beyond the total amount of funds in the entire health system. Thus, each provider competes for a bigger share of a fixed pie.

NEXT STEPS

Efforts to introduce planned market models in European health systems are now approaching the five year mark, and an increasing number of both official and informal evaluations have begun to appear (Jonsson, 1993; Bergman, 1994; Glennerster et al., 1994; Ham, 1994; Robinson and Le Grand, 1994; Saltman and von Otter, 1995). The technical issues in conducting adequate evaluations remain formidable. Among other complications, many recent health system reforms were implemented during the 1991–93 recession. As a consequence, it can be difficult to discern the specific impact of the reform measures alone. In some countries, the reform movement has been blamed for institutional cutbacks and job losses that appear to be more a function of hard economic times than the reform measures actually implemented.

As these evaluations are analysed, it may become possible to begin to provide answers to a variety of important health reform questions: Does increased efficiency harm solidarity? Do negotiated contracts cost more money than they save? Can patients make an appropriate decision about where to receive care? Can primary care providers utilize competitive incentives to gain greater leverage over hospitals? How do different planned market models influence health gain?

Once tentative answers to these and other essential questions start to emerge, it will become possible to draw more specific judgements about the relative impact and usefulness of particular planned market models. One key issue in the current economic and political climate will be the impact of various planned market approaches on overall expenditures, and on managerial concerns about value for money in the design and delivery of health services. An equally important question will be the differential ability of alternative planned market models to increase the effectiveness of the medical system as well as to generate health gain among the population as a whole. Answers to these and other difficult questions will be keenly awaited by all students of health system behaviour.

NOTE

1. The following section draws upon Saltman, 1995.

REFERENCES

Bergman, S.-E. (1994). *Likt et Olikt*. Stockholm: Spri.

Diderichsen, F. (1992). Market reforms in Swedish health care: a threat to or salvation for the universalistic welfare state? *International Journal of Health Services*, **23**, 185–8.

Friedman, M. (1962). *Capitalism and Freedom*. Chicago, University of Chicago Press

Glennerster, H., Matsaganis, M., Owens, P. and Hancock, S. (1994). *Implementing Fundholding: Wild Card or Winning Hand?*. Milton Keynes: Open University Press.

Ham, C. (1992). Managed competition in the NHS: progress and prospects. Paper delivered to the Manchester Statistical Society, 24 March.

Ham, C. (1994). *Management and Competition in The New NHS*. Oxford: Radcliffe Medical Press.

Hamel, G. and Prahalad, C.K. (1994). *Competing for the Future*. Boston: Harvard Business School Press.

Hammer and Champy, (1993). *Re-Engineering The Corporation*. New York: HarperCollins.

Jonsson, E. (1993). *Konkurrens inom sjukvården: Vad säger förskningen om effekterna?* Stockholm: Spri.

Polanyi, K. (1944). *The Great Transformation*. New York. Rhinehardt.

Porter, M. (1980). *Competitive Strategy*. New York: Free Press.

Rehnberg, C. (1995). The Swedish experience with internal markets In: M. Jerome-Forget, J. White and J.M. Weiner (Eds), *Health Care Reform Through Internal Markets: Experience and Proposals*. Brookings Institute, Washington, pp. 49–74.

Robinson, R. and Le Grand, J. (Eds) (1994). *Evaluating the NHS Reforms*. London, King's Fund Institute.

Saltman, R.B. (1994a). Patient choice and patient empowerment in northern European health systems: a conceptual framework. *International Journal of Health Services*, **14**, 201–29.

Saltman, R.B. (1994b). A conceptual overview of recent health care reforms. *European Journal of Public Health*, **4**, 287–93.

Saltman, R.B. (1995). *The Public–Private Mix in Financing and Producing Health Services*. Report to Population, Health, and Nutrition, World Bank, Washington, mimeo.

Saltman, R.B. and von Otter, C. (1989). Public competition versus mixed markets: an analytic comparison. *Health Policy*, **11**, 43–55.

Saltman, R.B. and von Otter, C. (1992). *Planned Markets and Public Competition: Strategic Reform in Northern European Health Systems*. State of Health Series. London: Open University Press.

Saltman, R.B. and von Otter, C. (Eds) (1995). *Implementing Planned Markets in Health Care: Balancing Social and Economic Responsibility*. State of Health Series. Buckingham, UK: Open University Press.

Smith, A. (1759). *The Theory of Moral Sentiments*. D. Raphael and A. Macfie (Eds), Clarendon Press, Oxford (1976).

von Otter, C. (1991). The application of market principles to health care. In: D. Hunter (Ed.), *Paradoxes of Competition for Health*. Nuffield Institute, University of Leeds.

von Otter, C. (1994). Reform Strategies in the Swedish Public Sector. In: Frieder Naschold and Marga Pröhl (Eds), *Produktivität öffentlicher Dienstleistungen*. Verlag Bertelsmann Stiftung.

Williamson, O. (1975). *Markets and Hierarchies: Analysis and Antitrust Implications*. New York: Free Press.

Chapter 2

Priority Setting: Political Issues

Chris Ham

INTRODUCTION

In this chapter I will explore the impact of fixed budgets on rationing or priority setting – terms which will be used interchangeably – in health care. The purpose of this chapter is to review approaches to priority setting in a number of systems in which budgets are fixed and to identify the lessons which emerge (Ham, 1995).

THE BACKGROUND

In a landmark study published in 1974, Maxwell analysed international trends in health and health services. Maxwell's study reviewed the pattern of expenditure on health care in different countries and the pressures to increase spending. This was not the first analysis of its kind but at a time when the oil crisis was forcing a reappraisal of public expenditure priorities on a worldwide basis it served to highlight the dilemmas facing policy makers. Since then the existence of infinite demand and finite resources in the health sector has become part of the conventional wisdom among health policy analysts.

If health care were left to the market, rationing decisions would be based on the ability of individuals to pay for services, or at least to insure against sickness. Given the well-known weaknesses of markets in health care, in practice governments in many countries have chosen to intervene to guarantee at least a minimum level of security against sickness. This means that responsibility for priority setting is shared between different stakeholders, most obviously politicians, insurers or purchasers, providers and patients.

Following Klein (1993), it can be suggested that priorities are in fact set at a number of levels. At the macro or systems level, there is the decision on the level of funding to be allocated to health services – the priority, that is, for health care in relation to other competing claims on resources. Next there are choices on the distribution of the

Fixing Health Budgets: Experience from Europe and North America.
Edited by F. W. Schwartz, H. Glennerster and R. B. Saltman. Technical editor R. Busse.

health care budget between different geographical areas and services, the subject of other chapters in this book. At a third level, decisions have to be made on the allocation of resources to particular forms of treatment within service areas – for example, the priority to be attached to hip replacements or knee replacements within orthopaedics. Then there are choices about how to prioritize access to treatment between different patients if treatment cannot be administered immediately. Finally, there are decisions on how much to invest in individual patients when access to care has been secured.

Choices at the macro or systems level are usually the province of politicians who, in health systems where the bulk of funding comes from public sources, take a close interest in the level of resources allocated to health care. In contrast, choices at the micro level are primarily the responsibility of doctors and other health care professionals. In between these two extremes a number of players are involved in decision making. The way in which decisions on rationing are made is not easy to understand. This is because these decisions are arrived at in the privacy of doctors' consulting rooms or in the committee rooms occupied by those responsible for policy making and resource allocation. The lack of transparency in rationing decisions presents a major problem for analysts seeking to understand what lies behind these decisions.

It is also the case that since rationing decisions are by definition almost always unpopular, there is a marked tendency for responsibility to be transferred from one group of stakeholders to another. To give an example, politicians in the UK, when confronted with evidence that certain services are under resourced, have been known to argue that this is the responsibility of local health authorities whose job it is to determine priorities for their areas within the budgets they are allocated. Similarly, doctors in the National Health Service (NHS) may seek to blame politicians when they are unable to provide services to the level they would wish. In this respect rationing is a political "hot potato" that no group wants to hold on to for too long.

INTERNATIONAL EXPERIENCE

All the more interesting (and surprising) therefore that politicians in a number of systems have started to address rationing decisions explicitly. The system that has attracted greatest interest is the state of Oregon which for the past five years has been seeking to define a basic package of care for the Medicaid programme (Kitzhaber, 1993). The intention of the initiators of the Oregon scheme was that the package of care agreed for Medicaid would also serve as the benchmark for employer-based insurance in the state.

After a lengthy period of development the Oregon scheme came into operation in 1994. In the version that has been implemented, 565 treatments out of a total of 696 have been funded. This has enabled low income groups previously excluded from Medicaid to be incorporated in the scheme in what has turned out to be a direct trade-off between the range of services funded under Medicaid and the extent of population coverage. A notable feature of the Oregon initiative was the effort made to involve the public in the setting of priorities.

A different approach has been taken in New Zealand where the government set up a core services committee in 1992 (Cumming, 1994). At an early stage in its work, the committee concluded that it was neither helpful nor sensible to seek to draw up a tightly defined list of services – Oregon-style – to be provided within the health care system. This view was based on the belief that the services actually provided represented the result of a series of decisions made over many years and as such were probably as good an assessment as could be made of the scope of service coverage.

Accordingly the committee went on to make a broad judgement of relative priorities within the existing range of services provided. It concentrated its efforts on advising on priorities within specific areas of service provision. This was achieved by staging consensus development conferences and using the results to formulate guidance on levels of service provision. This was linked to work focusing on the development of clinical guidelines to determine which patients should receive access to care.

In the Netherlands priority setting has focused on the work of a government-appointed committee (the Dunning Committee) set up to advise on the benefits package to be included in the reformed social insurance system. The committee reported in 1992 and put forward a framework to guide decision making. In brief, the committee recommended that services or treatments should be required to pass four tests before being included in the benefits package. These tests involved demonstrating that care was necessary from the community's point of view, that it was effective and efficient, and that it could not be left to personal responsibility. As in New Zealand, this was linked to a proposal that clinical guidelines should be developed to ensure that services were provided to patients who were able to benefit from treatment, and that explicit criteria should be used to determine the access of patients on to waiting lists and from waiting lists into hospital.

The UK government has so far not followed the example of the other countries whose experience has been reviewed here, and established a national committee or commission to advise on priorities. Instead, the government has argued that responsibility for priority setting is essentially a local matter. This means that health authorities have the task of determining priorities for their areas within the policy framework set at a national level. The experience of health authorities has been reviewed by an all-party parliamentary committee in a thorough inquiry into priority setting (House of Commons Health Committee, 1995).

The policy framework established by the UK government includes a long list of national objectives established over a period of years by politicians. As a consequence, health authorities suffer from priority overload. This means in practice that they are expected to deliver a wide range of goals and targets encompassing improvements in health, reductions in waiting times for hospital treatment, increases in efficiency and so on. While there has been some attempt to introduce greater coherence in the recent past, it remains the case that politicians are promiscuous when it comes to setting priorities and find it difficult to impose a self-denying ordinance.

One of the more positive features of the UK approach is the establishment of a research and development programme within the NHS. This has the aim of promoting knowledge-based decision making and evidence-based medicine. Of particular note is the work of the Cochrane Centre and the NHS Centre for Reviews and Dissemination. These organizations provide advice and information on cost-effectiveness in health care and this is intended to support the decisions taken by health authorities in the priority-setting process. Not least, the research and development programme provides health care purchasers with the ammunition to fire in their debates with providers about the allocation of scarce resources.

POLITICS AND TECHNIQUES

Viewed from a political perspective, an important feature of the debate about rationing is the attempt to turn questions about values into technical matters. This is perhaps most apparent in the pre-eminent position of economists in discussions of priority setting. At one level this is neither surprising nor undesirable as economics has much to contribute to decisions about health care priorities. Both in relation to assessing the cost-effectiveness of specific technologies and in comparing different technologies, there is an extensive literature that can contribute much to informed decision making. It is therefore entirely proper that the advice and input of economists has been actively sought by politicians seeking a route through the priority setting maze.

Yet as experience in Oregon has clearly demonstrated, priority setting cannot be reduced to a quick technical fix. For one thing there remain major gaps in our knowledge of cost-effectiveness of medical interventions. This led to some odd results (to say the least) when policy makers in Oregon invoked economics techniques in ranking different services. For another thing, approaches such as quality adjusted life years (QALY) and cost per QALY league tables include assumptions about values that may not be apparent to the uninitiated. Used without caution they give decision making a spurious objectivity which belies the basis on which they are constructed.

It is for this reason that commentators such as Klein (1992) have argued that techniques need to be used alongside debate in the priority-setting process. This is not to deny that economics and other disciplines have a part to play but rather is to make the point that information and evidence must be joined with argument as priorities are discussed. This process should be as open and pluralistic as possible to ensure that the needs of vulnerable groups are not neglected. And again to echo the Oregon experience, in seeking to involve the public in debate about priorities it is important that the discussion is not skewed in favour of community activists or those who through their employment in the health sector have a vested interest in the outcome.

To argue in this way is to suggest that priority setting needs to come out into the open. If rationing decisions cannot be justified on technical grounds – because there is no "right answer" to the question, what should be our priorities? – it is important that the way in which they are arrived at is transparent and defensible. This will become even more important in the future as the gap between demand and supply in health care widens. Unless politicians loosen the purse strings and increase the rate of growth of health spending, an option that is not impossible but at present seems unlikely, then choices in health care will be yet more difficult than they are today. In these circumstances, there seems little alternative to a fundamental re-examination of the scope of service provision in which as wide a set of interests as possible is involved.

MUDDLING THROUGH?

Yet while this is the route taken in both Oregon and The Netherlands, it has not so far found favour elsewhere. Indeed, as New Zealand discovered, the science of muddling through is deeply ingrained in the psyche of policy makers and policy advisers, suggesting that fundamental questions are easier to avoid than to address head on. This certainly applies in the UK, a system in which the myth of a health service offering services from the cradle to the grave has prevailed since 1948, and where there are simply no votes available to the brave (but possibly foolish) politician who unveils the myth and opens up a debate about the consequences.

The difficulty in not being brave is that rationing decisions continue to be taken implicitly and without any inherent logic. As an example, adult dental services have been withdrawn progressively from the NHS following a dispute about remuneration between the government and the dental profession. Much more seriously, there has been a reduction in the provision of continuing care within the NHS resulting in gaps in service provision on what appears to be a significant scale. This has affected elderly people and their carers particularly and has caused distress and dissatisfaction among groups who relied on the NHS to look after them in times of need but who are now beginning to realize that this will not necessarily happen.

The incremental responses of politicians to these developments bears all the hallmarks of the boy putting his fingers in the dyke. As a strategy this may be sufficient in the short term, but before long it will fail unless a more radical approach is taken. The flaw in this argument is of course that politicians – including health ministers – operate on short-term time-scales and work on the assumption that they will hold another portfolio by the time the land becomes inundated. The historical record shows that this is a reasonable assumption, but short termism has never been a good basis for constructing robust health policies.

The serious point here is that the postwar march to universal and comprehensive health care provision in almost all developed countries will increasingly be difficult to sustain. Without exploring the reasons in detail, this is because of the pressures that arise from the ageing population and expanding medical technologies. Coupled with what appears to be an almost universal desire to restrict public spending and reduce the tax burden on both individuals and employers, health systems funded principally out of public sources will come under increasing strain. If this assessment is correct, then something will have to give. This could either be universality or comprehensiveness. Conceivably it could be both.

THE CHALLENGE TO COMPREHENSIVENESS

From a political perspective it seems more likely that comprehensiveness will be sacrificed in order to at least maintain access to a basic benefits package for the entire population. The question then becomes the scope of the benefits package. Oregon is the only system to have made significant progress in this direction because, as we noted above, it was able to trade off restrictions in the availability of services for an increase in the groups covered. Those countries that have achieved universality are in a more difficult position and this helps to explain why they have so far chosen a different path.

This suggests that for the foreseeable future at least rationing decisions will continue to be characterized by muddling through. Indeed, there are those who maintain that this is in any case the best strategy in circumstances in which both economic analysis and approaches to involving the public are in their infancy (Hunter, 1993). Yet if it is accepted that the challenge of rationing is likely to become more difficult, then muddling through, even elegantly, will not be sufficient. A more coherent strategy is required including at a minimum a continuing and real commitment to health technology assessment, a set of incentives (economic, political and managerial) to encourage the adoption of effective technologies and to discourage the use of those that are ineffective, and an emphasis on the use of clinical guidelines and protocols to ensure that effective technologies are delivered appropriately.

The strategy would also encourage public participation in decisions on priorities and transparency in the decision-making process. There is an urgent need to promote greater health literacy – to borrow a phrase from a recent Secretary of State for Health in the UK – by involving the public in the choices that have to be made. This should help to promote more informed use of services by patients and their families, thereby reducing some of the inappropriate demands that exist. And at the macro level it will prepare the ground for the fundamental debate about comprehensiveness which I have argued in this chapter cannot be avoided for much longer.

CONCLUSION

Finally, it goes without saying that there is enormous scope for countries to learn from each other in this process. But much more needs to be done to create both formal and informal mechanisms for this to happen. In the UK there is already in place a learning network of organizations actively involved in work on priority setting co-ordinated by the University of Birmingham and an embryonic international network linking the UK and Sweden. Now is the time for other interested parties to join in.

REFERENCES

Cumming, J. (1994). Core services and priority-setting: the New Zealand experience. *Health Policy*, **29**, 41–60.

Ham, C.J. (1995). Synthesis: What can we learn from international experience? *British Medical Bulletin*, **51**, 319–30.

House of Commons Health Committee (1995). *Priority Setting in the NHS: Purchasing. First Report*, Volumes 1 & 2. London: HMSO.

Hunter, D. (1993). *Rationing Dilemmas in Health Care*. Birmingham: National Association of Health Authorities and Trusts.

Kitzhaber, J. (1993). Prioritising health services in an era of limits: the Oregon experience. *British Medical Journal*, **307**, 373–7.

Klein, R. (1992). Warning signals from Oregon. *British Medical Journal*, **309**, 1457–8.

Klein, R. (1993). Dimensions of rationing: who should do what? *British Medical Journal*, **307**, 309–11.

Maxwell, R. (1974). *Health Care The Growing Dilemma*. New York: McKinsey.

Chapter 3

Priority Setting: the Oregon Example

Theodore G. Ganiats
Robert M. Kaplan

HEALTH CARE PAYMENT IN THE UNITED STATES

The USA has a complex health care system. Some people are insured for medical expenses through private insurance companies. An estimated 41 million Americans have no health insurance at all (Office of Technology Assessment, 1992), and the remainder are insured through government programs. Medicare is a program that provides health insurance for all Americans older than 65 years. Medicaid is a program administered by each US state to provide health care for the poor, the blind, and the disabled. Medicaid expenses are divided between states and the federal government, with each paying 50% of the program costs. In 1990 Oregon had about 2.8 million residents. Among these, an estimated 150 000 received Medicaid (5%). It has been estimated that between 400 000 and 450 000 Oregon citizens (16% of population) are currently uninsured.

In recent years, the proportion of the poor who are covered by Medicaid has increased. The costs of the programs per capita have also increased, causing a financial crisis. Medicaid programs have responded by restricting the eligibility criteria. For most American states, the eligibility threshold for Medicaid is 50% of the federal poverty level. In current terms, that would make a family of three with an annual income of about $6500 too rich to be entered into the Medicaid program.

Further, Medicaid recognizes only specific categories. Those who are blind and disabled are eligible. However, poverty alone is not a sufficient condition to get help. For low income people to gain entrance into Medicaid, they must be a family with dependent children. A man or woman without children could not gain entrance to the program unless he or she were blind, disabled, or pregnant (if female).

The problems in Oregon were similar to most US states. In the late 1980s, the costs of the Oregon Medicaid program grew at a rate of 25% per year. In response to financial

Fixing Health Budgets: Experience from Europe and North America.
Edited by F. W. Schwartz, H. Glennerster and R. B. Saltman. Technical editor R. Busse.

pressures, the eligibility criteria were revised. In other words, people were being rationed rather than services. With the state unable to afford many basic health services, there was some concern about whether the underfunded public program should be paying for very expensive organ transplantation procedures. In 1987 a young boy in Oregon developed acute leukemia and his physicians decided that he needed a bone marrow transplant. At that time, the Oregon Medicaid program did not reimburse for transplants. The state legislature recognized that 34 transplants to Medicaid patients during the period 1987–1989 used the same financial resources as prenatal care and delivery to 1500 pregnant women. The legislature decided that they would use their limited resources to provide a small benefit to the large number of pregnant women instead of providing a large benefit to a small number of beneficiaries. The family of the young leukemia sufferer protested this policy, but the boy died while the case was under appeal.

The Oregon experience is not unique. Health care costs in the United States have risen rapidly over the last several decades. It is estimated that the United States will spend over one trillion dollars on health care in 1995. Despite substantial changes in the nature of health care in the United States, health care spending is still growing 50% faster than the gross domestic product (Burner, Waldo and McKusick, 1992). Other countries are experiencing similar difficulties in constraining health care costs (Schieber, Poullier and Greenwald, 1994). If resources were abundant, this may not be a problem. However, resources are limited, and these changes require that we use our resources wisely. Only recently have policy makers begun to recognize that not all health care expenditures are of equal value. Indeed, many resources are spent on health care services that provide little or no benefit. Conversely, many services that may be of high value cannot be supported from public resources.

As a result of these pressures, there is growing interest in finding models that will balance costs and benefits. Specifically, there is an intense need to find "value for money." In this chapter we consider one application of cost-utility analysis in the public sector. It is important to recognize that these methods may also be of interest to the private sector. Hospitals, insurers, and health maintenance organizations (HMOs) may also have interest in determining how to produce the most health given their resources.

This chapter presents the history of the Oregon experiment to move about 120 000 of the uninsured into Medicaid. First, we briefly discuss the problem valuing health services. Second, we review a general health policy model that has been used for health allocation decisions. Third, we review one application of this model in state of Oregon. Lastly, we discuss limitations of the model and outline a program of research that will be required in order to make productive use of decision-making models in the future.

VALUING HEALTH SERVICES

The rapid increase in health care costs has prompted cost cutting strategies. Often, cost reduction is considered the major objective of health care reform. Pauly (1995), for example, argues that cost should be the central consideration in policy analysis. However, too much attention on cost may neglect the mission of health care. For example, if cost is the only criterion, the development of guidelines for appropriate care may exclude expensive services. In order to choose between alternative health programs, it is best to evaluate not only the costs but the benefits (Sturm and Wells, 1995). Such an evaluation recognizes all financial and health outcomes as either a cost or a benefit. Financial outcomes are easily understood, but clinical outcome measures are often poorly understood, especially from a patient perspective. For example a change in an arterial blood gas value is not an ideal health outcome measure because it may not mean much to a patient or to a public policy maker. On the other hand, restoration or preservation of the ability to perform activities of daily living is the goal of many therapies. Patient-centered outcomes are measurable, and a paradigm shift in medicine is beginning to embrace patient-centered reports.

Despite the improvements in measuring patient outcomes, determining the value of health services has been particularly difficult. In contrast to cost-benefit analysis, which focuses on the dollar returns for investing in particular programs, cost-utility analysis used in some health services evaluations considers the health outcomes of a program, weighted by patient preferences for outcomes, in relation to the financial costs of the program. In economics the value of a product is related to the willingness of consumers to pay for it. For example, the value of a Mercedes Benz automobile is set by the price consumers are willing to pay for the car. If the price is too high, few cars are sold. Health services are difficult to value in this manner because consumers rarely pay for them directly. Instead, the charges are paid by third parties. Third party payment leaves consumers out of the loop and makes it difficult to establish whether the services are valuable to patients.

Another complication is that health care is a heterogeneous mix of services, yet all providers compete for the same resources. In order to make appropriate decisions, it is necessary to measure the value of all services using some common yardstick. In the next section, we present a model that can be used for these comparisons.

A GENERAL HEALTH POLICY MODEL

In order to understand health outcomes, it is necessary to build a comprehensive theoretical model of health status. This model includes several components. The major aspects of the model include mortality (death) and morbidity (health-related quality of life). In several papers, we have suggested that diseases and disabilities are important for two reasons. First, illness may cause the life expectancy to be shortened. Second, illness may make life less desirable at times prior to death – diminished health-related quality of life – (Kaplan, Anderson and Ganiats, 1993; Kaplan and Anderson, 1988).

Over the last two decades a group of investigators at the University of California, San Diego, has developed the General Health Policy Model (GHPM). Central to the model is a general conceptualization of quality of life. The model separates aspects of health status and life quality into distinct components. These are life expectancy (mortality), functioning and symptoms (morbidity), preference for observed health states (utility), and duration of stay in health states (prognosis).

Mortality. A model of health outcomes necessarily includes a component for mortality. Indeed, many public health statistics focus exclusively on this through estimations of crude mortality rates, age-adjusted mortality rates, and infant mortality rates.

Morbidity. Health-related quality of life is also an important outcome. Most public health indicators are relatively insensitive to variations toward the well end of the continuum. Measures of mortality, to give an extreme example, ignore all variations of morbidity: a person in a coma is considered equivalent to an asymptomatic person at full function. In addition, disability measures often ignore those who are relatively healthy. For example, the RAND Health Insurance Study reported that about 80% of the general population has no dysfunction. Thus, they would estimate that 80% of the population is well. In studies where symptoms and function are assessed, only about 12% of the general population report no symptoms on a particular day (Kaplan, Bush and Berry, 1976). In other words, health symptoms or problems are a very common aspect of the human experience. Some might argue that symptoms are unimportant because they are subjective and unobservable. However, symptoms are highly correlated with the demand for medical services, expenditures on health care, and motivations to alter life-styles. Thus, we feel that the quantification of symptoms is very important when assessing morbidity. The GHPM, using the Quality of Well-being (QWB) scale mentioned later, considers functioning in three areas (mobility, physical activity, and social activity) and symptoms.

Utility (Relative Importance). Not all outcomes are equally important. For example, a treatment in which 20 of 100 patients die is not equivalent to one in which 20 of 100 patients develop nausea. Given that mortality and the various components of morbidity can be tabulated, it is important to consider their relative importance. A key component of the GHPM attempts to scale the various health outcomes according to their relative importance. This exercise adds the "quality" dimensions to health status. In the preceding example the relative importance of dying would be weighted more than developing nausea. The weighting is accomplished by rating all states on a quality continuum ranging from 0 (for dead) to 1.0 (for optimum, asymptomatic functioning). These ratings are typically provided by independent judges who are representative of the general population. Using this system it is possible to express the relative importance of states in relation to the life–death continuum. A point halfway on the scale (0.5) is regarded as halfway between optimum function and death. The quality of life weighting system for the QWB has been described in several different publications (Kaplan, Bush and Berry, 1976, 1978, 1979).

Prognosis. Another dimension of health status is the duration of a condition. A headache that lasts one hour is not equivalent to a headache that lasts one month. A cough that lasts three days is not equivalent to a cough that lasts three years. In considering the severity of illness, duration of the problem is central. As basic as this concept is, most contemporary models of health outcome measurement completely disregard the duration component. In the GHPM the term "prognosis" refers to the probability of transition among health states over the course of time. In addition to consideration of duration of problems, the model considers the point at which the problem begins. A person may have no symptoms or dysfunctions currently but may have a high probability of health problems in the future. The prognosis component of the model takes these transitions into consideration. A discount rate is used for future outcomes if the utility of a future outcome is not the same as that of a present outcome. For example, a headache that will begin a year from now may be less of a concern than a headache that will start immediately.

QUALITY OF WELL-BEING SCALE

The Quality of Well-being (QWB) scale is one of several different approaches for valuing health outcomes (Kaplan, Anderson and Ganiats, 1993). Using this method patients are classified according to symptoms and distinct levels of functioning. These levels are represented by scales of mobility, physical activity, and social activity. Individuals are also classified by the symptom or problem that bothered them most. Symptoms or problems may be severe, such as serious chest pain, or minor, such as taking medication or a prescribed diet for health reasons. Human value studies have been conducted to place the observable states of health and functioning on to the 0 to 1.0 utility continuum mentioned above. Using these weights, the point-in-time QWB score can be calculated:

$$\text{QWB} = 1.0 - (\text{observed morbidity} \times \text{morbidity weight}) - (\text{observed physical activity} \times \text{physical activity weight}) - (\text{observed social activity} \times \text{social activity weight}) - (\text{observed symptom/problem} \times \text{symptom/problem weight})$$

The GHPM has been used in a wide variety of population studies (Anderson, Kaplan and DeBon, 1989; Erickson et al., 1989). In addition, the QWB scale has been used in clinical trials and studies to evaluate therapeutic interventions in a wide range of medical and surgical conditions. These include chronic obstructive pulmonary disease (Kaplan, Aitkins and Timms, 1984), AIDS (Kaplan et al., 1995), cystic fibrosis (Orenstein, Ross and Kaplan, 1989), diabetes mellitus (Kaplan et al., 1987), atrial fibrillation (Ganiats, 1993), lung transplantation (Squier et al., 1994), arthritis (Kaplan, Kozin and Anderson, 1988; Bombardier et al., 1986), cancer (Kaplan, 1993a), Alzheimer's disease (Kerner, Patterson and Kaplan, 1994), sinus disease (Hodgkin, 1994), and a wide variety of other conditions (Kaplan, 1993b). Specific validity and reliability data are available in each disease area. In cystic fibrosis patients, for example, QWB scores have been shown to be positively correlated with pulmonary function and exercise tolerance and negatively correlated with age. Further, the QWB was able to track improvement over time in a two-week intervention for treatment of pulmonary exacerbation, with changes in QWB statistically significantly correlated with changes in pulmonary function (Orenstein et al., 1989; Orenstein et al., 1990).

One can integrate components of the model to express outcomes in a terms of quality-adjusted life years (QALYs). A QALY is defined as the equivalent of a completely well year of life, or a year of life free of any symptoms, problems, or health-related disabilities. A principle advantage of the QALY is that it provides a common metric that allows different programs to be directly compared. The quality-adjusted life expectancy is the current life expectancy adjusted for diminished quality of life associated with dysfunctional states and the duration of stay in each state.

Consider, for example, a person who has rare lung disease who is in a state of functioning and symptoms that is rated by community peers as 0.5 on the 0 to 1.0 utility scale mentioned above. If the person remains in that state for one year, he or she would have lost the equivalent of 1/2 of one year of life. However, a person who has the flu may also be rated as 0.5. In this case, the illness might last only three days and the total loss in QALYs might be $3/365 \times 0.5$ which is equal to 0.004 QALYs. By itself it is clear that the flu does not produce as significant a health outcome as the lung disease. But suppose that 5000 people in a community get the flu. The QALYs lost would then be 5000×0.004 which is equal to 20 years, or greater than the one person with the rare lung disease. This indicates that the flu may be a greater health policy problem than the rare disease.

Now suppose that a vaccination becomes available and that the threat of the flu can be eliminated by vaccinating the 35 000 people in the community. The cost of the vaccine is \$5.00 per person or \$175 000. The average cost/utility of the program would be:

$$\frac{\text{net cost}}{\text{net utility}} = \frac{\$175\,000}{20 \text{ QALYs}} = \$8750/\text{QALY} \tag{3.1}$$

Using the concept of the QALY, the net cost/utility ratio of two programs can be calculated by:

$$\frac{\text{net cost}}{\text{net QALY}} = \frac{\text{net cost of treatment 2} - \text{net cost of treatment 1}}{[\text{QALY 2} - \text{QALY 1}]} \tag{3.2}$$

Table 3.1 summarizes the cost/utility for several programs that have been analyzed with the system.

Table 3.1 Summary of Cost/QALY estimates for selected medical, surgical, and preventive interventions[a]

Program	Reference	Cost/QALY
Seat belt laws	(Kaplan and Anderson, 1988)	0
Ante-partum and anti-D injection[b]	(Torrance and Zipursky, 1984)	1 543
Pneumonococcal vaccine for the elderly	(Office of Technology Assessment, 1979)	1 765
Coronary artery bypass surgery for left main disease	(Weinstein and Stason, 1982)	4 922
CABG (surgery) 2-vessel disease[f]	(Weinstein and Stason, 1982)	39 770
CABG (surgery) 1-vessel heart disease[f]	(Weinstein and Stason, 1982)	662 835
Neonatal intensive care, 500–999 g	(Boyle et al., 1983)	38 531
Neonatal intensive care, 1000–1499 g	(Boyle et al., 1983)	5 473
T4 (thyroid) screening	(Epstein et al., 1981)	7 595
PKU screening[c]	(Bush, Chen and Patrick, 1973)	8 498
Treatment of severe hypertension (diastolic > 105 mm Hg) in males age 40	(Weinstein and Stason, 1977)	10 896
Oral gold in rheumatoid arthritis	(Thompson et al., 1988)	12 059
Dapsone for prophylaxis for PCP pneumonia[d]	(Freedberg et al., 1991)	13 400
Treatment of mild hypertension (diastolic 95–104 mm Hg) in males age 40	(Weinstein, Stason and Blumenthal, 1976)	22 197
Oat bran for high cholesterol	(Kinosian and Eisenberg, 1988)	22 910
Cholestipol for high cholesterol	(Kinosian and Eisenberg, 1988)	92 467
Cholestyramine for high cholesterol	(Kinosian and Eisenberg, 1988)	153 105
Rehabilitation in COPD[e]	(Toevs, Kaplan and Aitkins, 1984)	28 320
Estrogen therapy for postmenopausal symptoms in women without a prior hysterectomy	(Weinstein et al., 1980)	32 057
Hospital hemodialysis	(Churchill, Lemon and Torrance, 1984)	40 200
Continuous ambulatory peritoneal dialysis	(Churchill, Lemon and Torrance, 1984)	54 460
School tuberculin testing program	(Bush, Fanshel and Chen, 1972)	43 250
Screening mammography	(Eddy, 1989)	167 850
Total hip replacement	(Liang et al., 1986)	293 029
Aerosolized pentamidine for prophylaxis of PCP pneumonia[d]	(Freedberg et al. 1991)	756 000

a All estimates adjusted to 1991 US dollars; *b* treatment for Rh immunization; *c* PKU, phenylketonuria; *d* PCP, pneumocystic carinii pneumonia, *e* COPD, chronic obstructive pulmonary disease; *f* CABG, coronary artery bypass graft.

THE OREGON EXPERIMENT

The dilemma in Oregon presented earlier is similar to the problem in essentially all Western countries. The legislature was faced with the costs of health care expanding much more rapidly than the budgets for Medicaid. Collectively, Oregon citizens spent approximately US$6 billion on health care in 1989, which is about three times what they spent in state income taxes. Oregon also recognized that American health care is not a two-tiered system, but rather a three-tiered system. The three-tiered system includes people who have regular insurance and can pay for their care, people enrolled in Medicaid, and a growing third tier of people who had no health insurance at all. By 1993 this third tier represented about 450 000 citizens, or about one-fifth the population of the state. In addition, another 230 000 were underinsured, and the trend indicated the number of uninsured and underinsured was steadily increasing. In order to address the growing crisis of funding for health care, the only alternatives were to: (1) change eligibility criteria and remove some individuals from the Medicaid roles, (2) continue to provide care to a large number of people, but limit the care they receive, or (3) increase revenues (taxes) to pay for the current system.

Led by a grass roots citizens group known as Oregon Health Decisions, it was argued that Oregon, like other states and countries, was already rationing health care. The problem was that rationing was implicit and not open to public scrutiny. As described above, people were being rationed rather than services. In other words, many individuals in need of care received none because they were in the wrong category. Pregnant women, for example, were covered. Thus a young woman employed as an hourly worker may be ineligible for health care, but an unemployed woman would become eligible if she became pregnant. Thus, the system created incentives to become pregnant in order to have a regular source of health care. The system allowed health care under Medicaid for poor families with young children but disallowed coverage for poor families with older children.

These arguments caught the attention of John Kitzhaber, MD, the President of the State Senate. Under Kitzhaber's leadership, Oregon passed three pieces of legislation to attack this problem. In this chapter we focus most specifically on Senate Bill 27. This bill mandated that health services be prioritized using something similar to the General Health Policy Model. The purpose of the prioritization was to eliminate services that did not provide benefit. The process of creating the prioritized list was an extremely difficult one. The Commission began by attempting to create a prioritized list of all health services. However, it soon became apparent that this was a nearly impossible task. Thus, the Commissioners searched for a combination of conditions and treatments that could be lumped together. They refer to these as condition–treatment pairs. Examples of these condition–treatment pairs are shown in Table 3.2. For example, the problem of rectal prolapse is paired with the treatment partial colectomy, while osteoporosis is paired with medical therapy.

Table 3.2 Examples of condition–treatment pairs

Condition	Treatment
Rectal prolapse	Partial colectomy
Osteoporosis	Medical therapy
Ophthalmic injury	Closure
Obesity	Nutritional and life-style counseling

A Health Services Commission was created in order to develop the prioritized list. This Commission obtained several sources of information. First, they held public hearings to learn about preferences for medical care in the Oregon communities. These meetings helped clarify how citizens viewed medical services. Various approaches to care were rated and discussed. On the basis of 48 town meetings which were attended by more than 1000 people, 13 community values emerged. These values included prevention, cost effectiveness, quality of life, ability to function, length of life, and so on (see Appendix 1). The major lesson from the community meetings was that citizens wanted preventive services. Further, the people consistently stated that the state should forgo expensive heroic treatments for individuals or small groups in order to offer basic services for everyone. In order to pay for these basic services, it was necessary to reduce spending elsewhere, and it was therefore important to rank services according to their desirability given the community values. The Commission chose to evaluate services using the QWB scale and a modified GHPM.

Two factors influenced the Commission's decision to modify the GHPM. First, the Commission could not conduct clinical trials for each of the many condition–treatment pairs. Further, estimation of treatment benefit using the QWB cannot be left to lay persons. So, the Commission formed a medical committee that had expertise in essentially all specialty areas and nearly all of the major provider groups in the state. The Oregon citizens were concerned about using weights from California in assigning priorities in their state. Thus, 1001 Oregon citizens participated in a separate weighting experiment. The weights were obtained in a telephone survey that was conducted by Oregon State University.

MULTIPLE LISTS

The Oregon Health Services Commission created multiple versions of the prioritized list. The first list was created in 1990. This list rank ordered 1692 condition-treatment pairs. The Health Services Commission used rough estimates of QWB changes and cost to create this list. Although this list has been the target of extensive criticism, the Health Services Commission never released it publicly. While they were working on the list, they completed a first draft and that draft was discussed at a public meeting.

The list was put together in a hasty manner. As a result, there were some obvious problems. For instance, 314 different medical therapies were all valued at $98.51 and an additional 177 were all estimated to cost $2560.65 (Tengs et al., 1995). The last 193 pairs on the 1692-item list were ranked alphabetically according to diagnosis. In addition, there were many counter-intuitive orderings on the list. For example, treatment for thumb sucking was ranked higher than treatment for AIDS. Although the 1990 list was clearly flawed, it is also important to emphasize that it was never officially released, and it was created hastily by an inexperienced and overburdened staff.

Perhaps the most discussed list was the one released in 1991. This list included 709 items. The list was based on the QWB scale to estimate benefit and included estimates of costs. In the 1991 list, the system was reorganized according to three basic categories of care: essential, very important, and valuable to certain individuals. Within these major groupings there were 17 subcategories. Nine of the 17 subcategories were classified as essential (see Appendix 2). The Commission decided to place greatest emphasis on problems that were acute, fatal, and reversible. In these cases treatment prevents death and there is full recovery, e.g. appendectomy for appendicitis and non-surgical treatment for whooping cough. Other subcategories classified as essential included maternity care, treatment for conditions that prevent death but do not allow full recovery, and preventive care for children. Listed as very important were treatments for non-fatal conditions that would return the individual to a previous state of health. Also included in this category were acute non-fatal one-time treatments that might improve quality of life. These might be hip replacements and cornea transplants. At the bottom of the list were treatments for fatal or non-fatal conditions that did not improve quality of life or extend life. These might be progressive treatments for the end stages of diseases such as cancer and AIDS or care for conditions for which the treatments were known not to be effective. In the revised approach, the Commission decided to ignore cost information and to allow their own subjective judgements to influence the rankings on the list. Appendix 3 summarizes the conditions selected from the top of the list, the middle of the list, and the bottom of the list. Unfortunately, the final exercise in Oregon resulted in many deviations from the GHPM. However, the exercise demonstrates an attempt to resolve the health care crisis on the basis of health outcome.

FEDERAL DENIAL, REVISION, AND ACCEPTANCE OF THE OREGON PLAN

Although Oregon passed legislation to go forward with their experiment, their efforts required approval by the Department of Health and Human Services (DHHS). Medicaid is funded jointly by both the US states and the US federal government, and any alteration in standard policy requires approval by the federal government. When Oregon filed their petition for a waiver of traditional Medicaid policies, substantial public debate erupted. Although they had been promised a decision early in 1991, no formal action had been taken by the summer of 1992. Unexpectedly, DHHS rejected Oregon's application in August 1992, suggesting that the proposal violated the Americans with Disability Act. The particular concern was that the preference survey for the quality of life dimension of the QWB "quantified stereotypic assumptions about persons with disability." The department suggested that people with disabilities would be systematically undervalued by the system.

The DHHS analysis assumed that people with disabilities would be discriminated against because treatment could not improve their chronic problems. However, this analysis is incorrect. The effectiveness of a treatment is based on the estimated course of illness with and without treatment. Treatment that makes life last longer, even without improvements in quality of life, will produce substantial benefits. For example, consider a person who has a chronic condition that places him or her in a state at 0.6 on the 0 to 1.0 scale. Assume that treatment will maintain the person at this level while withdrawal of treatment will result in death. According to the QWB model, the treatment produces 0.6 (calculated as 0.60 – 0.00) QALYs each year the person maintained in this state. A treatment that produces 0.6 years of life each year it is administered is a very powerful treatment in comparison to alternatives. Clearly such important and powerful treatments would not be eliminated in the Oregon system. Targeted for elimination would be treatments that produce little or no benefit.

It is true that people with disabilities have lower QWB scores than people without disabilities. Low scores suggest that resources should be used to aid these individuals. A low QWB score is not a badge of disgrace, but rather an indication that health status is not optimal and an indication of where funds need to be targeted. In a later section of this chapter we will consider several of the ethical issues raised by the Oregon experiment.

When the Bush Administration declined the Medicaid waiver, Oregon created a third list and released it in 1992. The list included 688 condition–treatment pairs. In creating this list, the Commission first considered the probability that each condition–treatment pair would improve survival. Second, for pairs that were tied with regard to life saving, condition–treatment pairs were ranked according to their ability to return the patient to an asymptomatic state of health. The Commission was careful not to use the term "quality of life," since that was a stumbling block in the Bush Administration review. For those pairs that remained tied, conditions were evaluated according to their ability to bring a patient from a symptomatic to an asymptomatic state. Next, pairs remaining tied were rank ordered according to their average cost, with preference being given to the least expensive options. Fifth, pairs remaining tied were ranked alphabetically. In the sixth step, the Commissioners reviewed the list and readjusted rankings.

The list was submitted for review to the DHHS in March of 1993. By this time, the Administration in Washington had changed. The Clinton Administration accepted the proposal for a waiver, but required Oregon to make some minor adjustments. In particular, they were concerned about treatments that would take patients from symptomatic to asymptomatic states. Since people with disabilities cannot become asymptomatic, the Administration remained concerned that the procedure violated the Americans with Disability Act.

In response to these concerns, a fourth list was released in 1993. Essentially, the procedure was the same as the 1992 list except that steps 2 and 3 were eliminated. These were the steps concerned with life-saving interventions that would return a person to a symptomatic state of health and those that would move a patient from a symptomatic to an asymptomatic state. This fourth list was completed in April 1993 and the Oregon plan was approved. It was implemented in February of 1994.

TENGS' CRITICISM

One of the most important criticisms of the Oregon plan has been offered by Tengs and her associates at the Harvard School of Public Health (Tengs et al., 1995). These investigators used rank order correlational methods to compare the Oregon lists with cost-effectiveness analyses that have been published in the literature. They report that the 1990 list showed essentially no correlation with published cost-effectiveness analyses. The 1991 list was correlated 0.39 with published studies, while the correlation of the 1992 list was correlated 0.25. The correlation with the 1993 list was 0.24. Even before publication, this study has received widespread attention. However, there are significant problems with this analysis. Most importantly, the Tengs' analysis does not consider health-related quality of life. In fact, all of the analyses used in their evaluation were studies that valued the cost per year of life saved, and studies considering cost per quality-adjusted life year were systematically eliminated. Since one of the most important aspects of medical care is to improve functioning and the quality of life, this is a serious problem. The rationale for disregarding health-related quality of life was that too few studies have evaluated these outcomes.

It is of interest to note the 1992 list, the one closest to using the GHPM, showed the strongest correlation with the published literature. The correlation of 0.39 is statistically significant and substantial. It is important to emphasize that the Oregon project and the previously published analyses were conducted using very different methods. Further, the Tengs' analysis makes the assumptions that the rank ordering of previously published studies is correct and that failure to replicate this rank ordering must be incorrect. It is known that there is substantial variability in the estimates cost/effectiveness from published studies.

ETHICAL ISSUES RAISED BY THE OREGON EXPERIMENT

The Oregon experiment has been a target for many legal, social, and ethical critics. A few of these concerns will briefly be reviewed.

Community Preferences were Used to Make Decisions about Individual Cases

This issue has two components. First is the question of whether a group's preferences should be applied to an individual case. This is a common question that highlights the basic methodology of the process. While decision analysis can be used to assist an individual patient to make a decision in a particular case, such an analysis requires an in-depth understanding of all the possible treatment options and health outcomes. In practice this is not possible for every health care encounter. In addition, it is not clear that health policy should be based on what is best for each individual patient. Instead, many argue that health policy should be based on what is best for most patients in a given situation or what is best for the group as a whole. Thus, using community preferences to make decisions about individual cases is inappropriate. On the other hand, there is no problem with community preferences being used to make health policy decisions if preferences are stable across groups. That leads to the second issue.

Preferences of Non-Medicaid Recipients were Used to Prioritize Services for the Poor

One of the most robust findings relevant to the General Health Policy Model is that preferences do not differ greatly across social or demographic groups. This is illustrated in comparisons of arthritis patients to non-patients (Figure 3.1), men to women (Figure 3.2), and medically insured to uninsured (Figure 3.3). In each case, these groups rated the same case descriptions and the points on the graphs are mean ratings of a particular item obtained from the two groups. In all cases, there is a strong linear relationship between the preferences. Even using completely different methodologies, preferences obtained from citizens in Oregon are remarkably similar to those obtained from residents of California (Figure 3.4). These finding have been replicated in several other studies (Kaplan, 1994).

The Rule of Rescue

Hadorn expressed serious concerns about the original Oregon priority list (Hadorn, 1991; Hadorn and Brook, 1991). As an emergency room physician, Hadorn was also concerned because many technologies that improve quality of life were ranked ahead of those that rescue people from imminent death. The rule of rescue argues that there is a moral obligation to invest in rescue whenever saving a life is a possibility. Life-saving procedures should always rank above those that can be delayed. In other words, we should not spend money on prevention if it detracts from the opportunity to rescue, even if the rescue effort is highly likely to fail.

According to the rule of rescue, the rankings provided by the Oregon Health Services Commission did not make intuitive sense. However, a review of Hadorn's statement suggests that there were problems with his analysis. For example, Hadorn reviewed the original version of the list and pointed out that tooth capping produced the same expected benefit as surgery for ectopic pregnancy, even though surgery for ectopic pregnancy saves lives in almost all cases. Hadorn reasoned that these two were similar because tooth capping provides a small benefit at a small cost, while surgery for

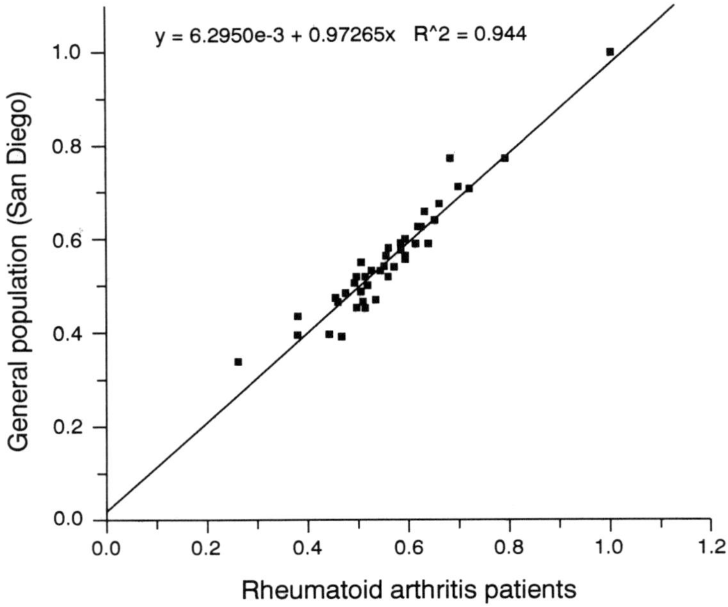

Figure 3.1 Comparison of rheumatoid arthritis patients versus general population (after Balaban et al., 1986)

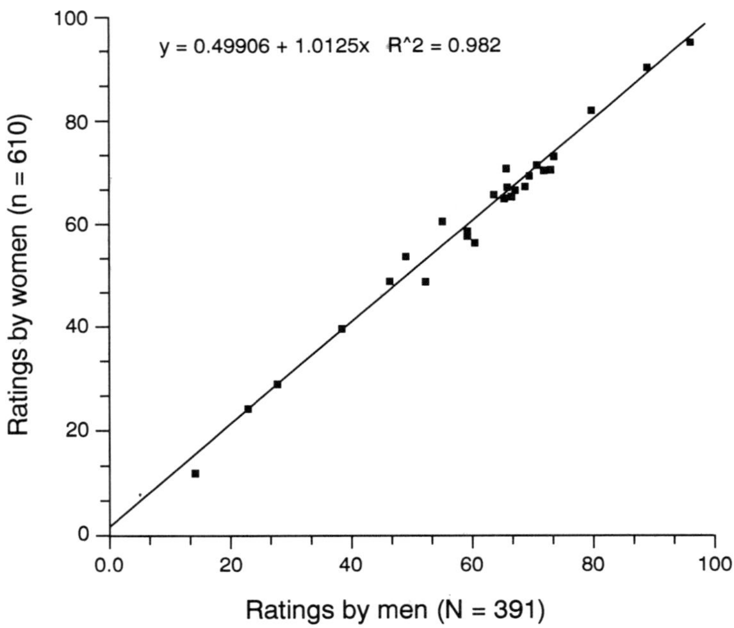

Figure 3.2 Comparison of case ratings by men and women in Oregon

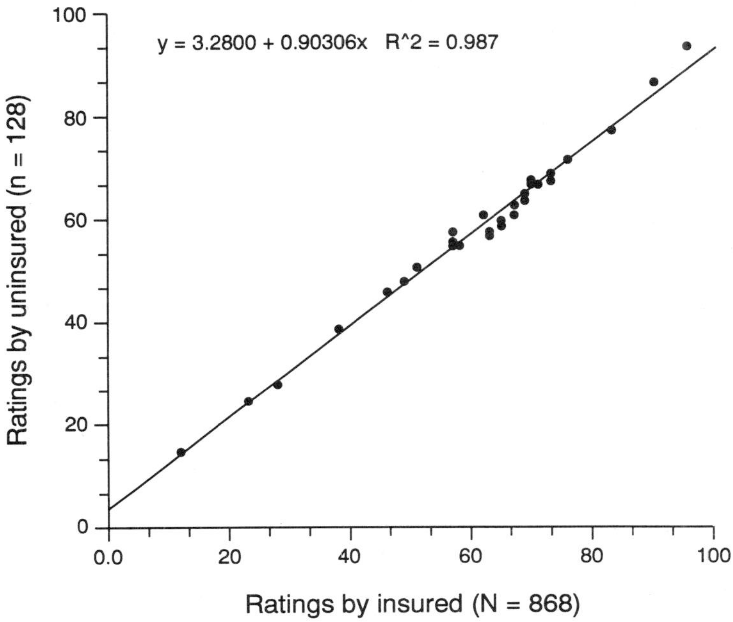

Figure 3.3 Ratings of 31 cases in Oregon by insured and uninsured citizens

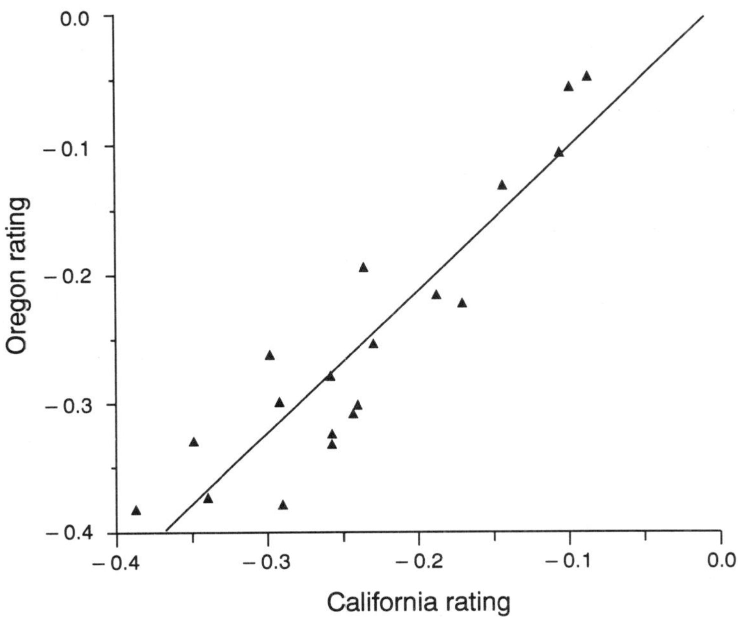

Figure 3.4 Scatterplot of California and Oregon ratings with three outliers removed

ectopic pregnancy provides a big benefit at a big cost. The cost/utility ratio appears to be the same, even though many would consider the management of ectopic pregnancy to be more important than capping a tooth.

There were several problems with Hadorn's analysis. In particular the cost of tooth capping was probably underestimated. He used a cost of $38, when the market price is over $500. Further, Hadorn overestimated the benefit of treatment for tooth capping by assuming that without treatment people would be in continuing dental pain and that this pain would be eliminated by the procedure.

Still, it is important to note that Hadorn identified that there was a substantial number of errors in the early construction of the list. Fortunately, the Oregon law allows continual revisions of the list when new data become available. The list that Hadorn referred to was quickly and appropriately abandoned by the Commissioners. Through successive review and critiques, the list will continually be improved.

The Program Discriminates Against Children

A consumer group known as The Children Defense Fund (CDF) attacked the Oregon program, asserting that it discriminated against children. CDF advocates programs for children through legal and political action. The major advocates are attorneys who expressed deep concerns about one essential component of the proposal. In order to have a prioritized list, it is necessary to offer malpractice immunity for physicians who deny some low priority services to patients. The CDF felt this was not acceptable.

In Oregon, poor women and children make up about 75% of Medicaid recipients. Further, the blind and the disabled were excluded from the initial version of the proposal. Critics charge that the proposal was a way of denying health care to those who needed care most. Although these arguments are compelling, it is also important to consider the benefits the program provides for poor women and children. Currently, thousands of women and children have no health care coverage at all. The Oregon plan makes funding available to an estimated 60 000 children and women who previously received no care. The GHPM in no way discriminates against children. Usually services for children are well evaluated because they have the potential to produce many years of benefit. In fact, critics have suggested that the model is biased in favor of children because lasting treatment effects are counted over and over again as the person ages. For example, prevention of a birth defect may produce large numbers of QALYs over the course of the life span.

Legal Challenges

The CDF vowed to launch several legal challenges against the Oregon proposal. In order to stop the experiment, the CDF argued that the Oregon plan creates a form of social experimentation in which women and children are forced to participate without consent. Further, they suggested that the program was unconstitutional because it discriminated on the basis of race and sex. The US Congress Office of Technology Assessment retained legal counsel to evaluate these challenges. The independent counsel suggested that the program did not violate equal protection clauses of the US Constitution, nor did it require the approval of a human subjects committee.

REMAINING PROBLEMS

This chapter has outlined a new paradigm for thinking about alternatives in health care. In short, it is suggested that limited health care resources be used to maximize health related quality of life. Services that do not work should not be funded and the savings should be used to extend basic health care benefits to people currently uninsured. The proposed model is consistent with the thinking of several groups, including scholars in the US (Office of Technology Assessment, 1979; Russell, 1986; Weinstein and Stason, 1977; Weinstein, Stason and Blumenthal, 1976), the UK (Drummond, Stoddard and Torrance, 1987; Maynard, 1991; Williams, 1988), Canada (Torrance, 1986, 1987), and Australia (Richardson, 1991). Although the exact methodologies proposed by these different research groups vary slightly, the theory is nearly identical. Recently, Patrick and Erickson (1993) offered a detailed account of the methodological steps required to implement the system. Methods are now available to begin guiding policy decisions. However, our information base for the implementation of the model is still incomplete.

Lack of Outcome Data

The attractiveness of the GHPM is based in several qualities. For example, it makes all choices explicit, so that anyone can evaluate the assumptions used to make decisions. The model focuses on key outcomes of particular concerns to health systems in the 1990s: quality of care, quality of life, mortality, and costs. In addition, the model appears to be quite robust, with flexibility to be used in a wide variety of health care systems.

The primary drawback to the model is a direct result of the first strength. By being explicit, the model requires a large amount of data, much more than what we have available for most health care programs. Little research has been completed that measures outcomes in a manner that is useful in the model. This requires that we either obtain the data directly (through clinical trials or other means) or that the data be estimated. Though there is extensive experience in using estimated data (Weinstein, Stason and Blumenthal, 1976; Naglie and Detsky, 1992), there is evidence that the methods used in this approach may be faulty (Fryback et al., 1993).

A second drawback is the effort required to complete an evaluation. A single analysis may take several months, and in order to evaluate properly all of the Oregon condition–treatment pairs would have been a formidable task. Work is needed on methods to simplify the evaluation without detracting from the strengths of the model.

Methodologic Problems

The Oregon plan posed several methodologic problems. For example, much work is needed to refine outcomes measurement models. In particular, methods for evaluating patient preferences (utility assessment) are fairly primitive. There are currently several ways to assess utilities in addition to the rating scales used in the QWB (e.g. time trade-off, standard gamble, willingness-to-pay), and often these methods do not arrive at the same utility weights. Which method is the most appropriate? Of course, the fundamental question relates to whether any kind of health policy analysis is appropriate. Perhaps all people should be offered care that meets a certain level of cost-effectiveness using their own preferences. As mentioned, there is evidence that mean preferences across groups are remarkably similar, but there is still wide variation within groups. It may be that any policy that uses group mean preferences actually does not maximize the outcome of the individuals in the group. Of course, any attempt to perform individualized analyses will run up against the tremendous resources required to perform a utility assessment with each patient.

More fundamental, the model assumes people act as predicted by Expected Utility Theory, but this is known to not be the case. Is this important? Should the model be modified to account for a different decision theory?

Ethical Concerns

Some authors have raised ethical concerns regarding the process of rationing. For example Brown (1991), in his essay on Oregon's plan, concluded, "American policymakers have not earned the right to ration health care, and the very policies that would earn it would eliminate much of the urge to exercise it." Etzioni (1991) proposed that before rationing health care, a government should look closely at limiting expenditures of much less merit, using space exploration as an example. These comments have gained both supporters and detractors.

Certainly, rationing occurs in all health care systems, though in the USA this has been described as the "implicit and invisible rationing of the current health system" (Fox and Leichter, 1991). The current system rations by denying coverage to some people rather than denying less effective services to all people. Indeed, two strengths of the Oregon system are its acknowledgment that the previous system rationed and the explicit nature of the new system's rationing methodology. Some have countered the proposal of postponing any rationing scheme until all system inefficiencies have been eliminated by claiming that any attempt to make the system more efficient would likely use many of the same techniques used in rationing: tough discipline, guidelines, denial of some forms of treatment believed efficacious by some, etc. (Callahan, 1991).

It is clear that both the dominant health care system in the USA and the experiment in Oregon have limitations. Under the dominant system health care costs have risen to the become the highest in the world, while 41 million citizens have no health insurance at all. It seems likely that any carefully designed social experiment that is supported by the people, such as the Oregon experiment, will give vital information on how to best mold the future of health care delivery in this country. The Oregon experiment is an attempt to follow principles of justice and to provide the most benefit for the most people. However the methodological and practical obstacles to implementation are substantial. The experience should help guide the future health care system in the USA.

Discounting

Psychologists have long recognized that people have a *time preference* for outcomes. Simply stated, time preference recognizes that most of us prefer a present reward over a future reward of equal value. Similarly, we usually prefer a future penalty over a present penalty of equal weight. Adjusting for this time preference is a standard element of current quantitative health policy analysis. This adjustment is called discounting (Weinstein et al., 1980)

A simple example illustrates discounting. As a lottery winner, would you choose a \$10 000 cash prize today or the same prize delivered in one year? Most of us would choose cash today. Even if the future award were adjusted for inflation (e.g. if inflation were 5%, the \$10 000 would grow to \$10 500), most of us would still prefer our winnings now. This choice implies that we value a future dollar less than a present one. In other words, we discount the value of future dollars.

People discount for a variety of reasons (Udry and Morris, 1971; Weinstein et al., 1980; Warner and Luce, 1982; Keeler and Cretin, 1983; Sackett, Haynes and Tugwell, 1985; Sox, Blatt and Higgins, 1988). First, they could invest the money and earn more than inflation. Second, they could spend money in hand and enjoy the benefits of the dollars now rather than in the future. Third, many do not trust the future. Some are not sure they will be alive in one year and want to enjoy the money now. Some suspect the source of the \$10 000 may default on the dollars owed.

The second and third factors indicate that our preference for today's dollar is not simply a matter of its investment potential; there other human factors at work. Thus discounting is based on multifactorial preferences for present over future dollars.

Economists, while comfortable with the concept of discounting, do not agree on what this discount rate should be. In cost-effectiveness research, proposed discount rates vary from 0% to 20% (Barnum, 1987; Grabowski and Hansen, 1990; Russell, 1987; Weinstein, 1980). Typically, analysts use a discount rate that approximates the prevailing interest earned on investments. Consideration of the above-noted human factors suggests that this is a low estimate.

In a quantitative policy analysis one usually assumes the economic and health discount rates are equal (Weinstein et al., 1980; Keeler and Cretin, 1983; Russell, 1987). The most often-quoted rationale for keeping the economic and health discount rates the same is the work of Keeler and Cretin (1983). They state that if the health discount rate is lower than the monetary discount rate, then no particular program should ever be implemented because the program will become more cost-effective by postponing it for one year. This is the result of the dollar numerator decreasing faster, under the influence of the larger discount rate, than the health outcome denominator: the calculated ratio must be smaller (i.e. more cost-effective) with each successive year. Others note that by keeping the monetary and health discount rates equal we assure the future relationship between the valuation of health and dollars remains constant and the same as today (Weinstein, 1980). Both of the arguments have been challenged (Ganiats, 1994).

Figure 3.5 shows the effect of discounting health outcomes over time and highlights another apparent paradox induced by current models. The current discounting model

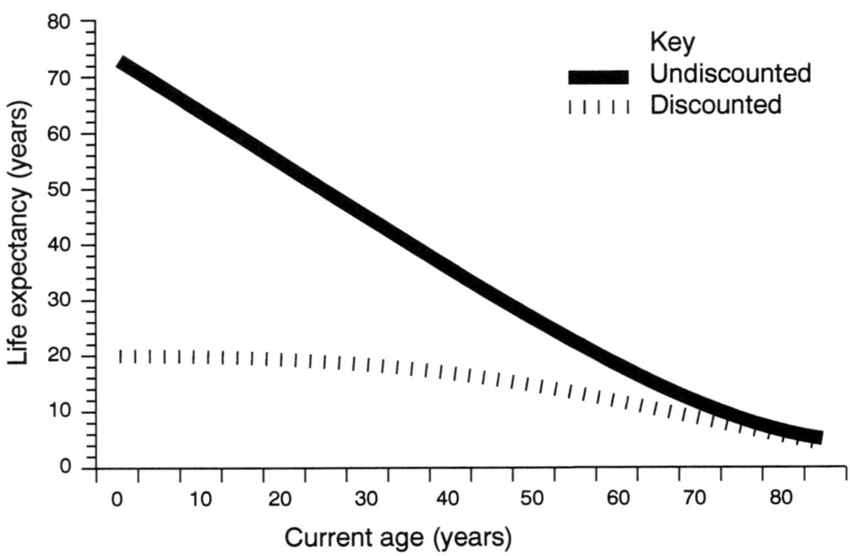

Figure 3.5 Example of a paradox: the effect of discounting life expectancy

tells us that the more distant in the future an outcome occurs, the less it is valued. Applying this premise to current life expectancy in the USA, we can see that the 73-year life expectancy of a newborn is valued as 20 years, while the 5.3-year life expectancy of an 85-year old is valued as 4.7 years. In other words, according to the current discounting model, the life expectancy of a newborn is about 15 years longer (20 years/5.3 years), or only about four times as long as the life expectancy of an 85-year-old.

Discounting can have such a major impact on the outcomes of an analysis, often the effect of discounting outweighs the effects of important components of the program itself. For this reason it is important for us to continue our research on how to best model the discounting of both economic and health outcomes. Until further research dictates a new direction, it is probably prudent to follow the standard course and discount both health and economic outcomes similarly.

CONCLUSION

Rapidly rising health care costs are forcing many to consider the value of the services provided. Rationing already takes place, but this is most often a rationing of people rather than a rationing of services. The US state of Oregon has recently implemented a program of health care resource allocation that attempts to provide the most efficient health services to the largest number of people. While there are several concerns regarding some of the theory and much of the method of this new program, it is a brave attempt to address this critical issue. Constant evaluation will enable the state to keep improving the program.

APPENDIX 1 HEALTH CARE VALUES ELICITED AS COMMUNITY MEETINGS

Prevention. Preventive services such as prenatal care and childhood immunizations were unanimously agreed upon as essential.

Quality of life. Services that enhance emotional and physical well-being as extend life, were generally thought to increase quality of life and should receive higher prioritization than those which only extended life.

Cost effectiveness. Cost-effective treatments were given high priority, although some community members disagreed that cost alone should be a primary determinant in prioritization.

Ability to function. The importance of independence and ability to perform daily activities was mentioned at three-fourths of the community meetings.

Equity. Equity was described as a fundamental belief that everyone should have equal assess to adequate health care. Discussions of equity raised various objections to the prioritization process. Many participants, for instance, thought that health care services should be available equally to all segments of society. There was support for increased federal funding for health care services, and some advocated establishment of a national health insurance plan. Other issues of equity included increasing access to treatment services on rural communities and universal access to health care for children.

Effectiveness of treatment. Treatments with proven efficacy and those that improve quality of life should be prioritized over those less likely to have successful outcomes.

Benefits many. Services that benefit many should receive higher priority than those from which few benefit.

Mental health and chemical dependency. Prevention, including drug education, was more highly valued than treatment services. While mental health and chemical dependency services were frequently discussed at meetings, there was some ambivalence regarding society's obligation to provide substance abuse services. Some participants, for example, felt that treatment was appropriate only in cases where patients were "motivated to undergo treatment," and that recidivism needed to be considered in cases of "repeat offenders."

Personal choice. Some community members expressed a desire for increased choice of type of providers, while others wanted more patient and family autonomy in making medical treatment decisions.

Community compassion. Society is obligated to provide treatments and services that alleviate pain and suffering (e.g. hospice care).

Impact on society. Treatment for infectious diseases and for alcoholism or drug abuse are examples of services that yield societal as well as individual benefit (discussed at approximately half of the community meetings).

Length of life. Prolonging life was viewed as the individual's obligation to society to seek appropriate health education and treatment services and generally to take responsibility for one's health. Individuals taking responsibility for their health should receive priority, and those whose illnesses are related to life-style, such as alcohol and drug-related conditions, are low priority if health care services are rationed.

APPENDIX 2 SUBCATEGORIES OF 1991 OREGON LIST

Every person is entitled to services necessary for a diagnosis. The following are the categories and some examples of services in each category.

Essential

1. *Acute fatal,* treatment prevents death and allows full recovery: appendectomy for appendicitis; non-surgical treatment for whooping cough; repair of deep, open wound in neck, non-surgical treatment for infection of the heart muscle (myocarditis).
2. *Maternity care,* including most newborn disorders: obstetrical care for pregnancy; care of the newborn.
3. *Acute fatal,* treatment prevents death but does not allow full recovery: non-surgical treatment for stroke; all treatment for burns; treatment for severe head injuries.
4. *Preventive care for children*: immunizations and well-child exams.
5. *Chronic fatal,* treatment improves life span and quality of life: non-surgical treatment for insulin dependent diabetes medical and surgical treatment for treatable cancer of the uterus; medical treatment for asthma; drug therapy for HIV disease.
6. *Reproductive services,* excludes maternity and infertility services; birth control and sterilization.

7. *Comfort care*: pain management and hospice care for the end stages of diseases such as cancer and AIDS.
8. *Preventive dental care*, adults and children: exams; cleaning and fluoride treatment.
9. *Proven effective preventive care for adults*: mammograms; blood pressure screening; Pap smears.

Very important

10. *Acute nonfatal*, treatment causes return to previous health: non-surgical treatment for acute thyroiditis; medical treatment for vaginitis; fillings for cavities.
11. *Chronic nonfatal*, one-time treatment improves quality of life: hip replacement; corneal transplants for cataracts; rheumatic fever.
12. *Acute nonfatal*, treatment without return to previous health: relocation of dislocated elbow, repair of cut to cornea.
13. *Chronic nonfatal*, repetitive treatment improves quality of life: non-surgical treatment for rheumatoid arthritis; gout; migraine headaches.

Valuable to certain individuals

14. *Acute nonfatal*, treatment speeds recovery: medical treatment for viral sore throat; diaper rash.
15. *Infertility services*: medical treatment for infertility; in-vitro fertilization; artificial insemination.
16. *Less effective preventive care for adults*: routine screening for those people not otherwise at risk, such as diabetes screening if the person is under 40 years old and not pregnant.
17. *Fatal or nonfatal*, treatment causes minimal or no improvement in quality of life: aggressive treatment for end stages of diseases such as cancer and AIDS; medical treatment for non-genital warts.

Each health service on the list is presumed to include necessary ancillary services such as hospital care, prescription drugs, and medical equipment and supplies necessary for successful treatment.

Source: Oregon Health Services Commission, in the public domain

APPENDIX 3 EXAMPLES OF CONDITION–TREATMENT PAIRS FROM TOP, MIDDLE, AND BOTTOM OF LIST

Top ten

1. Medical treatment for bacterial pneumonia
2. Medical treatment of tuberculosis
3. Medical or surgical treatment for peritonitis
4. Removal for foreign body from pharynx, larynx, trachea bronchus and esophagus
5. Appendectomy
6. Repair of ruptured intestine
7. Repair of hernia with obstruction and/or gangrene
8. Medical therapy for croup syndrome
9. Medical therapy for acute orbital cellulitis
10. Surgery for ectopic pregnancy

Middle ten

350. Repair of open wounds
351. Drainage and medical therapy for abscessed cysts of Bartholin's gland
352. Medical therapy for pilonidal cyst with abscess
353. Medical therapy for acute thyroiditis
354. Medical therapy for acute otitis media
355. PE Tubes or T and A for chronic otitis media
356. Surgical treatment for cholesteatoma
357. Medical therapy for sinusitis
358. Medical therapy for acute conjunctivitis
359. Medical therapy for spina bifida without hydrocephalus

Bottom ten

700. Mastopexy for gynecomastia
701. Medical and surgical therapy for cyst of the kidney
702. Medical therapy for end stage HIV disease (comfort care excluded – it is high on list)
703. Surgery for chronic pancreatitis
704. Medical therapy for superficial wounds without infection
705. Medical therapy for constitutional aplastic anemia
706. Surgical treatment for prolapsed urethral mucosa
707. Paracentesis of aqueous for central retinal artery occlusion
708. Life support for extremely low birth weight (500 g) and under 23 week gestation
709. Life support for anencephalous

REFERENCES

Anderson J.P., Kaplan R.M. and DeBon, M. (1989). Comparison of responses to similar questions in health surveys. In F. Fowler (Ed.), *Health Survey Research Methods*. Washington, DC: National Center For Health Statistics, pp. 13–21.

Barnum, H. (1987). Evaluating healthy days of life gained from health projects. *Social Science and Medicine*, **24**, 10, 833–41.

Bombardier, C., Ware, J., Russell, I.J., et al. (1986). Auranofin therapy and quality of life for patients with rheumatoid arthritis: results of a multicenter trial. *American Journal of Medicine*, **81**, 565–78.

Boyle, M.H., Torrance, G.W., Sinclair, J.C. and Horwood, S.P. (1983). Economic evaluation of neonatal intensive care of very-low-birth-weight infants. *New England Journal of Medicine*, **308**, 22, 1330–7.

Brown, L.D. (1991). The national politics of Oregon's rationing plan. *Health Affairs*, **10**, 28–51.

Burner, S.T., Waldo, D.R. and McKusick, D.R. (1992). National health expenditures projections through 2030. *Health Care Financing Review*, **14**, 1, 1–29.

Bush, J.W., Chen, M. and Patrick, D.L. (1973). Cost-effectiveness using a health status index: analysis of the New York State PKU screening program. In R. Berg (Ed.), *Health Status Indexes*. Chicago: Hospital Research and Educational Trust, pp. 172–208.

Bush, J.F., Fanshel, S. and Chen, M. (1972). Analysis of a tuberculin testing program using a health status index. *Journal of Social-Economic Planning Science*, **6**, 46–9.

Callahan, D. (1991). Ethics and priority setting in Oregon. *Health Affairs*, **10**, 78–87.

Churchill, D.N., Lemon, B.C. and Torrance, G.W. (1984). A cost-effectiveness analysis of continuous abulatory peritoneal dialysis and hospital hemodialysis. *Medical Decision Making*, **4**, 4, 489–500.

Drummond, M.F., Stoddard, G.L. and Torrance, G.W. (1987). *Methods for the Economic Evaluation of Health Care Programmes*. Oxford: Oxford University Press.

Eddy, D.M. (1989). Screening for breast cancer. *Annals of Internal Medicine*, **111**, 389–99.

Epstein, K.A., Schneiderman, L.J., Bush, J.W. and Zettner, W.R. (1981). The "abnormal" screening of Thyroxine (T4): analysis of physician response, outcome, cost and health effectiveness. *Journal of Chronic Disease*, **34**, 175–90.

Erickson, P., Kendall, E.A., Anderson, J.P. and Kaplan, R.M. (1989). Using composite health status measures to assess the nation's health. *Medical Care*, **27**, Supplement 3, S66–S76.

Etzioni, A. (1991). Health care rationing: a critical evaluation. *Health Affairs*, **10**, 88–95.

Fox, D.M. and Leichter, H.M. (1991). Rationing care in Oregon: the new accountability. *Health Affairs*, **10**, 7–27.

Freedberg, K.A., Tosteson, A.N., Cohen, C.J. and Cotton, D.J. (1991). Primary prophylaxis for pneumocystis carinii pneumonia in HIV-infected people with CD4 counts below 200/mm^3: a cost-effectiveness analysis. *Journal of Acquired Immune Deficiency Syndromes*, **4**, 5, 521–31.

Fryback, D.G., Dasbach, E.J., Klein, R., Klein, B.E., Dorn, N., Peterson, K., et al. (1993). The Beaver Dam health outcomes study: initial catalog of health-state quality factors. *Medical Decision Making*, **13**, 2, 89–102.

Ganiats, T.G. (1993). Are are outcomes created equal? *Family Practice Research Journal*, **13**, 1, 1–5.

Ganiats, T.G. (1994). Discounting in cost-effectiveness research. *Medical Decision Making*, **14**, 3, 298–300.

Grabowski, H.G. and Hansen, R.W. (1990). Economic scales and tests. In B. Spilker (Ed.), *Quality of Life Assessments in Clinical Trials*. New York: Raven Press, pp. 61–70.

Hadorn, D.C. (1991). Setting health care priorities in Oregon: cost-effectiveness meets the rule of rescue. *Journal of the American Medical Association*, **265**, 2218–25.

Hadorn, D.C. and Brook, R.H. (1991). The health care resource allocation debate. Refining our terms. *Journal of the American Medical Association*, **266**, 3328–31.

Hodgkin, P.S. (1994) Health impact of endoscopic sinus surgery assessed by the Quality of Well-being (QWB) Scale. Unpublished paper.

Kaplan, R.M. (1993a). Quality of life assessment for cost/utility studies in cancer. *Cancer Treatment Review*, **19**, Supplement A, 85–96.

Kaplan, R.M. (1993b). Application of a general health policy model in the American health care crisis. *Journal of the Royal Society of Medicine*, **86**, 277–81.

Kaplan, R.M. (1994). Value judgment in the Oregon Medicaid experiment. *Medical Care*, **32**, 10, 975–88.

Kaplan, R.M. and Anderson, J.P. (1988). A general health policy model: update and applications. *Health Services Research*, **23**, 203–35.

Kaplan, R.M., Anderson, J.P. and Ganiats, T.G. (1993). The Quality of Well-being scale: rationale for a single quality of life index. In S.R. Walker and R.M. Rosser (Eds), *Quality of Life Assessment: Key Issues in the 1990s*, 2nd edn. London: MTM Press, pp. 65–94.

Kaplan, R.M., Atkins, C.J. and Timms, R. (1984). Validity of a quality of well-being scale as an outcome measure in chronic obstructive pulmonary disease. *Journal of Chronic Diseases*, **37**, 2, 85–95.

Kaplan, R.M., Bush, J.W. and Berry, C.C. (1976). Health status: types of validity and the index of well-being. *Health Services Research*, **11**, 4, 478–507.

Kaplan, R.M., Bush, J.W. and Berry, C.C. (1978). The reliability, stability and generalizability of a health status index. In American Statistical Society (Ed.), *Proceedings of the Social Statistics Section*. Washington, DC: American Statistical Society.

Kaplan, R.M., Bush, J.W. and Berry, C.C. (1979). Health Status Index: category rating versus magnitude estimation for measuring levels of well-being. *Medical Care*, **17**, 5, 501–25.

Kaplan, R.M., Kozin, F. and Anderson, J.P. (1988). Measuring quality of life in arthritis patients (including discussion of a general health-decision model). *Quality of Life and Cardiovascular Care*, **4**, 131–9.

Kaplan, R.M., Hartwell, S.L., Wilson, D.K. and Wallace, J.P. (1987). Effects of diet and exercise interventions on control and quality of life in non-insulin-dependent diabetes mellitus. *Journal of General Internal Medicine*, **2**, 220–8.

Kaplan, R.M., Anderson, J.P., Patterson, T.L., McCutchan, J.A., Weinrich, J.D., Heaton, R.H., et al. (1995). Validity of the Quality of Well-being scale for persons with HIV infection. *Psychosomatic Medicine* (in press).

Keeler, E.B. and Cretin, S. (1983). Discounting of life-saving and other nonmonetary effects. *Management Science*, **29**, 300–6.

Kerner, D., Patterson, T.L. and Kaplan, R.M. (1994). Validity of the Quality of Well-being Scale in Alzheimer's disease. In preparation.

Kinosian, B.P. and Eisenberg, J.M. (1988). Cutting into cholesterol. Cost-effective alternatives for treating hypercholesterolemia. *Journal of the American Medical Association*, **259**, 15, 2249–54.

Liang, M.H., Cullen, K.E., Larson, M.G., et al. (1986). Cost-effectiveness of total joint arthroplasty in osteoarthritis. *Arthritis Rheumatism*, **29**, 937–43.

Maynard, A. (1991). Economic issues in HIV management. In A. Maynard (Ed.), *Economic Aspects of HIV Management*. London: Colwood House Medical Publications, pp. 6–12.

Naglie, I.G. and Detsky, A.S. (1992). Treatment of chronic nonvalvular atrial fibrillation in the elderly: a decision analysis. *Medical Decision Making*, **12**, 239–49.

Office of Technology Assessment (1979). *A Review of Selected Federal Vaccine and Immunization Policies: Based on Case Studies of Pneumococcal Vaccine*. Washington DC: Government Printing Office.

Office of Technology Assessment (1992). *Evaluation of the Oregon Medicaid Proposal*. Washington DC: Government Printing Office.

Orenstein, D.M., Nixon, P.A., Ross, E.A. and Kaplan, R.M. (1989). The quality of well-being in cystic fibrosis. *Chest*, **95**, 344–7.

Orenstein, D.M., Nixon, P.A., Ross, E.A. and Kaplan, R.M. (1990). Quality of well-being before and after antibiotic treatment of pulmonary exacerbation in cystic fibrosis. *Chest*, **98**, 1081–4.

Patrick, D.L. and Erickson, P. (1993). *Health Status and Health Policy: Allocating Resources to Health Care*. Boston, MA: Cambridge University Press.

Pauly, M.V. (1995). Valuing health care benefits in money terms. In F. Sloan (Ed.), *Valuing Health Care: Costs Benefits and Effectiveness of Pharmaceuticals and Other Medical Technologies*. New York: Cambridge, pp. 99–124.

Richardson, J. (1991). *Economic Assessment in Health Care: Theory and Practice*. Monash University: National Centre for Health Program Evaluation.

Russell, L.B. (1986). *Is Prevention Better than Cure?* Washington, DC: The Brookings Institution.

Russell, L.B. (1987). *Evaluating Preventive Care: Report on a Workshop*. Washington, DC: The Brookings Institution.

Sackett, D.L., Haynes, R.B. and Tugwell, P. (1985). Clinical Epidemiology – *A Basic Science for Clinical Medicine*. Boston: Little, Brown.

Schieber, G.J., Poullier, J.P. and Greenwald, L.M. (1994). Health system performance in OECD countries, 1980–1992. Organization for Economic Cooperation and Development. *Health Affairs*, **13**, 4, 100–12.

Sox, H.C., Blatt, M.A. and Higgins, M.C. (1988). *Medical Decision Making*. Boston: Butterworth.

Squier, H., Kaplan, R.M., Ries, A.L., et al. (1994). Quality of Well-being predicts survival in lung transplantation. Submitted for publication.

Sturm, R. and Wells, K.B. (1995). How can care for depression become more cost-effective?

Journal of the American Medical Association, **273**, 1, 51–8.

Tengs, T.L., Mayer, G., Saugel, J.E., Pliskin, J.S., Graham, J.D. and Weinstein, M.C. (1995). Oregon's Medicaid ranking and cost-effectiveness: is there a relationship? Unpublished manuscript.

Thompson, M.S., Read, J.L., Hutchings, H.C., Patterson, M. and Harris, E.D. (1988). The cost-effective of Auranofin: results of a randomized clinical trial. *Journal of Rheumatology*, **15**, 35–42.

Toevs, C.D., Kaplan, R.M. and Atkins, C.J. (1984). The costs and effects of behavioral programs for chronic obstructive pulmonary disease. *Medical Care*, **22**, 1088–100.

Torrance, G.W. (1986). Measurement of health state utilities for economic appraisal: a review. *Journal of Health Economics*, **5**, 1.

Torrance, G.W. (1987). Utility approach to measuring health-related quality of life. *Journal of Chronic Disease*, **40**, 593–600.

Torrance, G.W. and Zipursky, A. (1984). Cost-effectiveness of antepartum prevention of RH immunization. *Clinics in Perinatology*, **11**, 2, 267–8.

Udry, J.R. and Morris, N.M. (1971). A spoonful of sugar helps the medicine go down. *American Journal of Public Health*, **61**, 776–85.

Warner, K.E. and Luce, B.R. (1982). *Cost-benefit and Cost-effectiveness Analysis in Health Care: Principles, Practice, and Potential.* Ann Arbor, MI: Health Administration Press.

Weinstein, M., Stason, W. and Blumenthal, D. (1976). *Hypertension – A Policy Perspective.* Cambridge: Harvard University Press.

Weinstein, M.C. and Stason, W.B. (1977). Foundations of cost-effectiveness analysis for health and medical practice. *New England Journal of Medicine*, **296**, 716-21.

Weinstein, M.C. and Stason, W.B. (1982). Cost-effectiveness of coronary artery bypass surgery. *Circulation*, Supplement 3, 56–66.

Weinstein, M.C., Fineberg, H.V., Elstein, A.S., Frazier, H.S., Neuhauser, D., Neutra, R.R., et al. (1980). *Clinical Decision Analysis.* Philadelphia: W.B. Saunders Company.

Williams, A. (1988). The importance of quality of life in policy decisions. In S. Walker and R. Rosser (Eds), *Quality of Life: Assessment and Application.* London: MTP Press, pp. 279–90.

Chapter 4

Efficiency of Spending Under Fixed Budgets

Alan Maynard

INTRODUCTION

Health policy formulation, like clinical medicine, is rarely evaluated systematically and, as a consequence, the knowledge base is incomplete and, in some cases, contentious. Four economists visiting the People's Republic of China in 1993 agreed the policy advice set out in Appendix 1: major decisions about resource allocation in health care are not left to the market; markets are a *means* of achieving equity, efficiency and cost containment, not an *end* in themselves; the prevention of ill health requires strong government actions; and commercial insurance should be avoided, with the non-profit insurance industry being tightly regulated by itself and government.

Inherent in this advice was a belief that health care markets are always regulated and that an essential part of that regulation is fixed global budgets preferably funded by a "single pipe" of finance. The advocacy of global budgets has a long history in health policy (e.g. Reinhardt, 1978) and although the research base to substantiate such advocacy is limited (Wolfe and Moran, 1993) it does appear that this device may be a necessary but not a sufficient condition to ensure cost containment.

However, whether global budgets are an effective means of improving allocative efficiency is another matter. Countries with global budgets in health care (e.g. UK, Canada, Germany and New Zealand) exhibit familiar variations in clinical practice which are produced by the uncertainty inherent in physician decision making. This is induced both by the absence of a knowledge base for most health care interventions and a refusal to translate knowledge, where it exists, into practice - for example, many proven interventions in pregnancy and child birth, which are set out in the Cochrane database (Cochrane Pregnancy and Childcare Database, 1993), have not been applied locally. Thus while global budgets may contain costs they have, unless complemented by vigorous policies, little effect on the efficiency of resource allocation.

Fixing Health Budgets: Experience from Europe and North America.
Edited by F. W. Schwartz, H. Glennerster and R. B. Saltman. Technical editor R. Busse.

The purposes of this chapter are to review the use of global budgets in the UK's National Health Service (NHS) where, until recently, the policy focus has been on structure and finance, and then explore the nature and content of the policies supplementary to global budgets which may increase the efficiency with which scarce health care resources are used.

POLICY CHOICES: THE DOMINANCE OF COST CONTAINMENT OVER EFFICIENCY AND EQUITY

The roles of the Treasury (i.e. the Finance Ministry) and the professions (especially the medical and nursing professions) in health care policy formation in England are of great importance. The provider professions, in partnership with management groups, advocate more expenditure and this behaviour is partly altruistic and partly self-motivated. Their analyses typically focus on the demand pressures associated with the "greying" of the population and technological change and, in the UK, the uncritical assumption that at least 2% annual expenditure is needed to meet these increased demands. The evidence base for demographic and technological inflation is poor and it seems likely that the former can be met at modest cost even with low British gross domestic product growth levels and the latter may be inflationary because of inappropriate clinical practices and commercial pressures exerted in the absence of good cost-effectiveness data.

An alternative explanation of the advocacy of increased expenditure by the professions is that their behaviour is a manifestation of self-interest. As George Bernard Shaw argued, professions may be "a conspiracy against the laity", i.e. a self-interested group concerned more with their own income and welfare than with the interests of consumers and society. Reinhardt explained the expenditure equals health providers' income equation very nicely over a decade ago (Reinhardt, 1978) (Table 4.1). Thus

Table 4.1 A New Dictionary for Health Policy Analysts

Traditional terminology (based on the service-expenditure facet of the health care sector)	New terminology (based on the employment-income facet of the health care sector)
1. National health care expenditures	1. National health care incomes
2. The demand for health services	2. The supply of health care incomes
3. The supply of health manpower	3. The demand for health care incomes
4. Containment of health care expenditures	4. Containment of health care incomes
5. Health care cost containment expenditures	5. Containment of the income provided per unit of health service
6. Increased efficiency in the production of health services	6. Reduction in employment per unit of health service
7. Federal subsidies to health manpower training	7. Federally stimulated increases in the demand for health care incomes

Source: Reinhardt (1978)

containment of health care expenditure equals containment of health providers' incomes. Increased efficiency in the production of health care equals reduction in employment per unit of health service. Professional providers do not like reduced income or employment and hence they support expenditure growth (which equals cost inflation and enhancements in their material welfare) and may oppose both cost containment and efficiency policies.

This self-interest is traditionally countervailed by the Finance Ministry (in the UK, the Treasury) whose concern is macro economic stability. Thus the effects of the trade cycle, and the fluctuations in inflationary pressure which it brings, leads to periodic "squeezes" in expenditure, often focused on the capital programmes in services such as the NHS. However, this control is generally global and poorly targeted. In the 1980s in the UK, hospital expenditure was tightly controlled with policy emphasis and expenditure being relatively greater in community care and primary care (the latter being demand determined with an open ended budget).

During the 1980s expenditure control in the NHS was relatively tight and it was not until the internal market reforms of the NHS that expenditure rose rapidly. The Government has used its power as the funder of the NHS to squeeze the service with a programme of "efficiency gains" whereby hospitals were required to increase activity by increasing amounts over the last decade within fixed budgets. At present these "gains" are set at 3% per year and have been met each year despite providers' protests of the increasing difficulty of achieving them. The Treasury believe that increasing pressure of this nature can increase productivity (i.e. activity) further.

One way in which hospitals have met these efficiency gain targets is to squeeze some activity out of the NHS. This is particularly evident in the care of the elderly where social security expenditure on nursing homes for the elderly has risen from £700 000 in 1979 to over £2 billion in the 1990s. This public money finances private care outside the NHS.

The inflation in expenditure (Figure 4.1) is particularly marked after 1990 for the series adjusted by the NHS pay and prices deflator. These time series indicate that while global budgets may be a necessary condition for expenditure control, they are not sufficient. Where the political imperative is acute, governments may increase expenditure rapidly.

In the case of England that imperative was absent in the 1980s. The Thatcher administration regarded the NHS as an ideological anachronism suitable for privatization. A "think tank" report in 1981 made proposals which emphasized an increased role for private insurance: indeed one minister (Gerard Vaughan) predicted that by the end of the decade 25% of health spending would come from private insurance. However, the Cabinet was convinced otherwise by Norman Fowler who drew on researchers' work (e.g. McLachlan and Maynard, 1982) to demonstrate to his

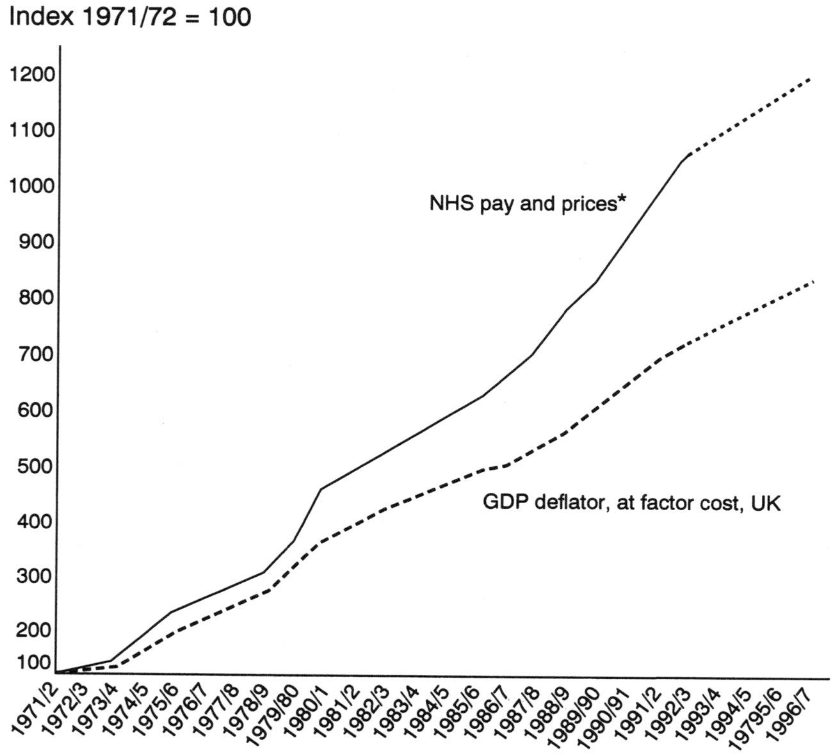

Source: OHE (1995, Figure 2.3)

Figure 4.1 UK total health care expenditure

colleagues and Prime Minister Thatcher that such privatization would not only lead to cost inflation by also to possibly increased inefficiency and inequity.

Instead of privatization, public expenditure controls were retained and structural reform was emphasized. The configuration of the NHS was changed in 1982 (with the abolition of Area Health Authorities) and management was reformed in 1984 with the abolition of consensus team management and its replacement with the Griffiths structure, with its emphasis on clearly identified responsibilities and short-term contracts for managers.

These reforms failed to remedy the inefficiencies in resource allocation in the NHS which, gradually over the 1980s, not only became evident to politicians but also became a source of acute frustration to them. The Adminstration was confronted by a regular cycle of media-induced hysteria about people being denied treatment and having to wait long periods for non-emergency elective procedures. Attempts to increase "throughput" (i.e. activity) to reduce waiting times and use resources more effectively were met by accusations from the professions of the "underfunding" of the NHS and a manifest reluctance to change practices and be more accountable. The "social contract" between the health care professions and Government was broken. After nearly 40 years during which Government had left the health professions to govern themselves and determine how resources were allocated in the NHS in exchange for the professions leaving the Government to determine the global budget, the "good and the great" of the medical profession met with Thatcher in late 1987 to tell her that the NHS was in crisis due to "underfunding". As a consequence the Prime Minister decided the medical profession and the NHS were suitable cases for "treatment"!

Mrs Thatcher established a review of the NHS early in 1988. Initially the review focused on the possibilities of private finance and the emulation of education reforms to turn hospitals into Trusts. A change of minister saw the latter proposal survive but the former was dismissed, as the focus of the review became supply side inefficiencies rather than demand side issues of "underfunding". Thus a process which was precipitated by political agitation about demand (finance) problems produced an outcome which maintained the global budget (and the potential for cost control) and began, very clumsily, to focus on supply side inefficiencies.

The internal market reforms, whilst declared a great "success" in the absence of systematic evaluative evidence by ministers, involved policy instruments such as:

1. The purchaser-provider split, where often the best management talent and clinical expertise was located in the provider, and purchasers with little evidence-based knowledge, often slight medical expertise and all too little managerial talent have produced much rhetoric but all too little of substance in shifting resources to more efficient uses.

2. The process of medical (now called clinical) audit with little management and "capture" by the medical profession has cost over £200 million and produced little benefit. The prospects of it being developed better now that its management has passed to the purchasers seem limited given the absence of relevant skills and knowledge. The recent decision of the Department of Health to create an Audit Review Centre to be run by the two provider trade unions (i.e. the British Medical Association and the Royal College of Nursing) is as supine as it is bizarre! (Department of Health, 1995.)

3. The development of general practice fundholding (GPFH) with "over generous" funding and little evaluation has had unknown effects. The number of quantitative studies of GPFH is small and their quality often low (Gosden and Torgerson, 1996 forthcoming): they indicate that after three years there are no substantive differences in prescribing and referral practices between fundholders and

non-fundholders. Qualitative and descriptive studies of GP Fundholding and primary care more generally (e.g. Glennerster et al., 1994; Leese and Bosanquet, 1995) provide indicators of apparent improvement but whether these are due to structural changes (e.g. the creation of GPFH) or increased resourcing is impossible to determine given the design (largely qualitative) of the studies.

4. Increased activity levels in NHS hospitals may be due to the reforms and/or increased funding. The extent to which this increased activity is efficient is unclear as there is no evaluation of the "appropriateness" and cost-effectiveness of interventions. Major variations in intervention and physical activity persist with increased managerial demands reducing the clinical time available for patient care perhaps to inappropriate (i.e. potentially dangerous) levels (e.g. Yates, 1995).

From an emphasis on cost containment with global budgets, an increasing awareness of the need to pursue efficiency goals has emerged in NHS policy. However, this is not the first time this concern has emerged as a central component of government policy; for example, the Labour administration 1974-79 was much concerned with both this and Health of the Nation targets (Department of Health and Social Security, 1976, 1977). Why is it that public policy makers have been unable to pursue effectively the more efficient use of resources? Why is it that researchers have been unable to persuade these decision makers to translate their findings into practice?

PURSUING EFFICIENCY MORE EFFECTIVELY

The impact of research and the efforts of researchers to disseminate their work is quite limited. Often this feature is attributed to the inadequacies of the researchers, i.e. they are accused of having failed to disseminate their findings. However, "it takes two to tango" - that is, in addition to there being a supply of knowledge, there has also to be a demand for it. Policy makers behave as if they are reluctant to be "confused" by facts, although, on occasion, they adopt research results such as those on the weighted capitation allocation formula, adapting them suitably to meet the vote maximization needs of their political masters (Peacock and Smith, 1995). The slow impact of research on policy formation and clinical practice may be a recognition of factors such as the primacy of the cost containment goal, an inability of policy makers to appraise and comprehend the content of the knowledge base, and the reluctance of politicians and policy makers to pursue efficiency when it involves the redistribution of providers' incomes and employment in ways which precipitate the loss of votes in the political marketplace. All too often it seems that when efficiency becomes a policy issue, it is a short-term interest which is applied with vigour only for the duration of the current cost containment crisis.

The lag between researcher advocacy and the adoption of research knowledge can be observed over many decades. In 1972 Archie Cochrane acknowledged the fact that most interventions by his fellow professionals had no knowledge base due to the failure of clinicians to evaluate scientifically the effectiveness of their work. He argued that the recognition of this problem would, over the next 20 years, be resolved. He wrote:

> Allocations of funds and facilities are nearly always based on the opinions of senior consultants, but, more and more, requests for additional facilities will have to be based on detailed argument with "hard evidence" as to the gain to be expected from the patients' angle and the cost. Few can possibly object to this. (Cochrane, 1972, p. 82)

So far this optimism has been unfounded. The 1990s has seen the Government's "discovery" of what Cochrane argued over 20 years ago, and the creation of the Cochrane Centre and the NHS Centre for Reviews and Dissemination (Sheldon and Chalmers, 1994). A similar "discovery" in the USA of the absence of a knowledge base to clinical practice led to the creation of the Patient Outcome Review Teams (PORTs).

This enthusiasm for research and development may be short term. The work of the PORTs is being criticized for methodological reasons (e.g. its dependency on time series analysis rather than the use of prospective randomized controlled trials (e.g. Sheldon, 1994) and due to the impatience of policy makers for "instant" results: perhaps too much was promised in too short a time period by the research community? Scepticism about the ability of the research community to deliver new knowledge quickly, together with a Republican dominated Congress anxious to cut expenditure, has led to the proposed abolition of the Office of Technology Assessment and to large cuts in the funding of the PORTs programme.

The rapid development of the NHS R & D programme in England (Department of Health, 1993) may also have difficulty meeting the high expectations of potential users in Whitehall and the health service. The shortage of high-quality researchers to implement the programme is evident and the use of the large Centres model which will form the core of the R & D activity is unproven, risky and experimental. The ability of the Cochrane collaboration (an international network of review activities across different therapeutic and care fields) to review effectiveness, let alone the ability of groups such as the NHS Centre for Reviews and Dissemination to disseminate such evidence and translate this knowledge base into changed behaviour, is limited due to funding constraints and the complexity of the tasks, e.g. the scientific review of the effectiveness is complex and will take many years (or even decades) and the knowledge base about how to change doctors' behaviour is slight.

There is also the risk that such activity will be "captured" by narrow clinical interests motivated by job status preservation at worst and the definition of effectiveness at best. As long ago as the early 1970s, Alan Williams (1972, 1974) argued in favour of the economic evaluation of health care interventions and produced guidelines for the efficient practice of these arts. The similarities between his guidelines and the advisory set now agreed between the pharmaceutical industry and the British Government as a voluntary code can be seen from Appendix 2 (A and B). Progress has been so slow in translating this researcher advocacy into policy practice: the usual two decade lag as evident here as it is in relation to Cochrane's (1972) advocacy of systematic appraisal.

This emphasis on economic evaluation is at variance with the emphasis of some elements of the R & D strategy which emphasizes effectiveness alone. The Cochrane Centre's focus is on effectiveness as were the initial publications in the Effective Health Care Bulletin series (e.g. those on lipid lowering drugs and depression). This is unfortunate. For instance, intensive community care for schizophrenics may be more effective than inpatient care. However, the latter is cheap and the former is expensive: resource allocation choices have to be based on cost-effectiveness not effectiveness data alone.

The reluctance of some clinical researchers to include economic variables in their trials (e.g. the large trial of the prophylactive use of tamoxifen ignores economic issues) is indicative both of a reluctance to acknowledge that such data are relevant and criticism of the methods of economic evaluation. Clinical researchers accuse economic researchers of defects in the design and execution of their work. Without doubt, both clinical and economic evaluations, are often deficient, sometimes biased and occasionally fraudulent. Altman (1994) has argued that the primary purpose of much clinical research is the development of curriculum vitae and enhancement of promotion and career, rather than the production of new knowledge and better patient care.

Much economic evaluation is built on clinical research of uncertain quality. Such modelling is difficult and may produce poor results. For instance, economic modelling based on clinical trials of non steroidal anti-inflammatory drugs (NSAIDs) may be difficult due to the corruption of many of these trials (Gotzsche, 1989). Similar problems may exist with economic modelling of alternative forms of anti-depression drug therapy (Freemantle and Maynard, 1994).

The practice of economic evaluation exhibits failures to practice the principles of good practice elucidated over the last two decades (Appendix 2). Reviews of practice have detailed the defects in the practice of economic evaluation (e.g. Uhvarhelyi et al., 1992). Systematic appraisals of the stock and flow of such studies (which is now underway in the Centre for Health Economics and the Centre for Reviews and Dissemination at the University of York) may identify deficient practice and deficient practitioners, although whether this information will be sufficient to alter practice is not known. However, the results of trials of the effectiveness of intervention are of limited use if those making choices about resource allocation do not also have the relevant cost data. Without such data inefficiencies may be continued and scarce resources will be wasted by inefficient and unethical clinical practices.

The obstacles to the translation of knowledge about efficient practice into more efficient behaviour are numerous:

1. Politicians, policy makers, purchasers and providers have recognized the need to appraise and improve efficiency in the past but their attention to such issues has been spasmodic, superficial and uneven.

2. The task of identifying "what works" in medicine is complex, costly and time intensive. International standards (criteria) for reviewing effectiveness and cost-effectiveness are being developed and, when applied, will identify those cost-effective therapies which need to be applied, those (many) therapies which are unproven, and those therapies which are inefficient. The identification of unproven therapies will help formulate research strategies internationally. The identification of proven therapies will provide a challenge for those seeking to translate knowledge into practice. However this resource intensive research enterprise will take many years to yield information to inform medical practices about what is and is not cost-effective.

3. The knowledge base is absent to inform the process of ensuring that, with contracting, education and other processes, proven (i.e. cost-effective) therapies are implemented. What are the relative advantages of alternative pecuniary and non-pecuniary incentive systems in altering provider behaviour?

4. Concentration on effectiveness rather than cost-effectiveness will not facilitate the achievement of greater efficiency in resource allocation, e.g. a policy to eradicate H-pylori as a source of stomach ulceration with antacids and antibiotics appears to be effective and cost effective. Other policies may be effective but not cost-effective. Choices should not be determined with only one half of the relevant data set, however much clinicians like to focus on effectiveness and ignore costs!

5. The tasks of identifying and using proven therapies and proven means of changing professional behaviour is an international challenge. The need for good trials (i.e. with large numbers of patients and professionals) means that multi-centre international trials will be unavoidable. No nation (not even Britain) can be an island, isolated from the worldwide need to determine the efficiency of resource allocation in health care. Billions of pounds, dollars and marks can be saved and patient welfare improved if research is funded efficiently and patiently, i.e. with recognition that the dividends of research will take many years to achieve. If appropriate resources are not made available, the debate started in academia decades ago and in currency with many governments now will lead to little change and the "re-invention" of these issues in the next century.

6. The processes of translating knowledge into practice requires clear identification of the relative policy making roles of the central government department, purchasers, providers, professional organizations and other agencies. Central co-ordination of

R & D, to avoid duplication, seems sensible. Purchasers have to transform themselves from "price takers" (i.e. passive recipients of funding who act as bank clerks who pay out cash) to "price makers", who bear down on providers, demanding evidence of cost structures and quality data and working with them to write cost-effectiveness evidence into contracts. There is the risk (as seen perhaps in the USA) that such managed care purchasers may focus unduly on costs and produce quality deterioration. Such risks have to be managed with care: a process facilitated if appropriate cost-effectiveness data are produced. Finally health care professionals need to be trained in the techniques of systematic appraisal of data so that the degree to which they are simultaneously informed and "bamboozled" is changed to facilitate efficient patient care (Sackett and Rosenberg, 1995).

CONCLUSION

In 1732 Francis Clifton, the physician to a Hanoverian King of England argued:

> In order, therefore to procure this valuable collection, I humbly propose, first of all, that three or four persons should be employed in the hospitals (and that without any ways interfering with the gentlemen now concerned), to set down the cases of the patients there from day to day, candidly and judiciously, without any regard to private opinions or public systems, and at the year's end publish these facts just as they are, leaving every one to make the best use he can for himself. (Francis Clifton, 1732)

Today, in the UK, Germany and all other countries this advice is ignored.

In 1747 James Lind conducted a "randomized controlled trial" giving fresh citrus fruit to the crew of one of his ships and none to the other two boats. When they arrived in India, scurvy had affected the crew of the latter severely but not the crew of the former. This discovery did not affect policy for nearly 50 years: in 1795 the British navy made lime juice mandatory on all its boats.

Given the advocacy of Clifton and the "science" of Lind, which was ignored for nearly 50 years, why should we be more optimistic in the 1990s? If the objective of society is the efficient use of scarce resources, policy makers, politicians and the societies they represent should confront the first certainty of life - death - and manage it with the second certainty of life - the scarcity of resources - by "confusing" their choices with facts about the cost-effectiveness of competing policy options. Perhaps the best way to control costs within global budgets and move towards greater efficiency is to adopt the following motto when dealing with providers anxious to increase both their incomes and (one hopes) their capacity to produce health and social care for their patients:

> In God we trust!
> All others (particularly physicians) bring data!

REFERENCES

Altman, D.G. (1994). The scandal of poor medical research. *British Medical Journal*, **308**, 283-4.

Association of the British Pharmaceutical Industry and Department of Health (1994). Guidelines for the economic evaluation of pharmaceuticals. Press Release, 19 May.

Clifton, F. (1732). *An Essay on Physick*. London.

Cochrane, A.L. (1972). *Effectiveness and Efficiency: Random Reflections on Health Care*. London: Nuffield Provincial Hospitals Trust.

Cochrane Pregnancy and Childbirth Database (1993). Derived from the Cochrane Database of Systematic Reviews - Published through Cochrane Updates on Disk. Oxford: Update Software, Disk Issue 2.

Department of Health and Social Security (1976). *Priorities for Health*. London: HMSO.

Department of Health and Social Security (1977). *Prevention and Health*. London: HMSO.

Department of Health (1993). *Research for Health*. London: Department of Health.

Department of Health (1995). BMA and RCN win contract to run new national clinical audit information centre. Press release, 1 March.

Evans, R., Maynard, A., Preker, A. and Reinhardt, U. (1994). Health Care Reform. *Health Economics*, 3, 359-60.

Freemantle, N. and Maynard, A. (1994). Something rotten in the state of clinical and economic evaluations? *Health Economics*, **3**, 63-7.

Glennerster, H., Matsaganis, M., Owens, P. and Hancock, S. (1994). GP fundholding: wild card or winning hand? In R. Robinson and J. Le Grand (Eds), *Evaluating the NHS Reforms*. London: King's Fund.

Gosden, T. and Torgerson, D. (1996 forthcoming). *Prescribing and Referral Patterns: Does Fundholding Make a Difference?* National Primary Care Research and Development Centre at the Centre for Health Economics, University of York, mimeograph.

Gotzsche, P.C. (1989). Methodology and overt and hidden bias in reports of 196 double-blind trials of nonsteroidal anti-inflammatory drugs in rheumatoid arthritis. *Controlled Clinical Trials*, **10**, 31-56.

Leese, B. and Bosanquet, N. (1995). Family doctors and change in practice strategy since 1986. *British Medical Journal*, **310**, 705-8.

McLachlan, G. and Maynard, A. (Eds) (1982). *The Public Private Mix for Health*. London: Nuffield Provincial Hospitals Trust.

Office of Health Economics (1995). *Compendium of Health Statistics*. London: OHE.

Peacock, S. and Smith, P. (1995). The resource allocation consequences of the new NHS Needs formula. Discussion paper 134. Centre for Health Economics, University of York.

Reinhardt, U. (1978). Tablemanners at the health care feast. In D. Yaggy and W.A. Anylan (Eds), *Financing Health Care: Competition Versus Regulation*. Cambridge, MA: Ballinger.

Sackett, D.L. and Rosenberg, W.M.C. (1995). On the need for evidence-based medicine. *Health Economics*, **4**, 249-54.

Sheldon, T.S. (1994). Please bypass the PORT. *British Medical Journal*, **309**, 142-3.

Sheldon T.S. and Chalmers, I. (1994). The UK Cochrane Centre and the NHS Centre for Reviews and Dissemination: respective roles within the information systems strategy of the NHS R & D programmes, coordination and principles underlying collaboration. *Health Economics*, **3**, 201-3.

Uhvarhelyi, S., Colditz, G.A., Arti Rani, A.B. and Epstein, A.M. (1992). Cost-effectiveness and cost benefit analyses in the medical literature: Are the methods being used correctly? *Annals of Internal Medicine*, **116**, 238-44.

Williams, A. (1972). Cost-benefit analysis: bastard science? And/or insidious poison in the body politick? *Journal of Public Economics*, **1**, 199-225.

Williams, A. (1974). The cost benefit approach. *British Medical Bulletin*, **30**, 252-6.

Wolfe, P.R. and Moran, D.W. (1993). Global budgeting in OECD countries. *Health Care Financing*, **14**, 3, 55-76.
Yates, J. (1995). Dead in the water? *Health Services Journal*, 17 March, 21.

APPENDIX 1 CLOSING COMMUNIQUE DELIVERED TO THE SENIOR POLICY SEMINAR ON HEALTH FINANCING IN CHINA

by the following foreign experts:

Prof. Robert Evans, Univ. Of British Columbia, Canada
Prof. Alan Maynard, York University, England
Prof, Alexander Preker, The World Bank
Prof, Uwe Reinhardt, Princeton University, USA

July 31, 1993, Beijing, China

1. Established market economies do not use the market mechanism to govern their health care sectors. Market forces are enlisted here and there where it is safe to do so. But the important decisions regarding the allocation of resources are NOT left to the market in any Western country, even the United States.
2. "Competition" and "markets" should be means to an end, but not ends in themselves. In this seminar, "competition" and "markets" have often been treated as if they were ends in themselves. If they are treated as ends, the objectives of efficiency, equity and cost containment will NOT be achieved.
3. Better health is the major objective of health systems. Despite the great gains that China has made in achieving good health at low cost, the burden due to PREVENTABLE deaths and illnesses is still enormous. To remedy the situation and avoid losing previous gains, there is a continuing need for strong government actions.
4. One reform of the Chinese health system that would be most destructive of its proper objective would be to allow the development of *commercial* health insurance. Western countries that use health insurance to finance their health sectors do so either through public insurance or very tightly *regulated, non-commercial* social insurance. The only exception, the United States, is desperately trying to escape from the negative consequences of commercial insurance.

APPENDIX 2 GUIDELINES FOR ECONOMIC EVALUATION

(A)

1. What precisely is the question which the study was trying to answer?
2. What is the question that it has actually answered?
3. What are the assumed objectives of the activity studied?
4. By what measures are these represented?
5. How are they weighted?

6. Do they enable us to tell whether the objectives are being attained?
7. What range of options was considered?
8. What other options might there have been?
9. Were they rejected, or not considered, for good reasons?
10. Would their inclusion have been likely to change the results?
11. Is anyone who has not been considered in the analysis likely to be affected?
12. If so, why are they excluded?
13. Does the notion of cost go wider or deeper than the expenditure of the agency concerned?
14. If not, is it clear that these expenditure cover all the resources used and accurately represent their value if released for other uses?
15. If so, is the line drawn so as to include all potential beneficiaries and losers, and are resources costed at their value in their best alternative use?
16. Is the differential timing of the items in the streams of benefits and costs suitably taken care of (e.g. by discounting, and, if so, at what rate)?
17. Where there is uncertainty, or there are known margins of error, is it made clear how sensitive the outcome is to these elements?
18. Are the results, on balance, good enough for the job in hand?
19. Has anyone else done better?

Source: Williams (1974)

(B)

1. The question being addressed by the study, including the demographic characteristics of the target population group, should be identified and set out at the start of the report of the study.
2. The conceptual and practical reasons for choosing the comparator should be set out and justified in the report of the study.
3. The treatment paths of the options being compared should be identified, fully described, placed in the context of overall treatment and reported.
4. The perspective of the study should ideally be societal, identifying the impact on all parts of society, including patients, the NHS, other providers of care, and the wider economy. However, costs and outcomes should be reported in a disaggregated way so that the recipients of costs and outcomes can be identified. Attention should be drawn to any signficiant distributional implications.
5. The study should use a recognized technique. These include Cost Minimization Analysis (CMA), Cost Effectiveness Analysis (CEA), Cost Utility Analysis (CUA), and Cost Benefit Analysis (CBA). Any one of these could be appropriate according to the purpose of the study. The report of the study should include justification of the technique chosen.
6. In choosing the method of data capture and analysis, the use of one of, or a combination of, prospective or retrospective randomized clinical trials, meta-analysis, observational data and modelling should be considered. The reasons for choice of method and, where relevant, for choice of trials should be reported.
7. Assessment of the question should include determiing and reporting what additional benefit is being provided at what extra cost using incremental analysis of costs and outcomes.
8. Outcome measures should be identified and the basis for their selection reported. Where CUA is used, proven generic measures of Quality of Life are preferred.

9. All relevant costs should be identified, collected and reported. Physical units of resource use should be collected and reported separately from information about the costs of the resources. Costs should reflect full opportunity cost, including the cost of capital and administrative and support costs where relevant. Average cost data is often acceptable as a proxy for long-run marginal cost.
10. Discounting should be undertaken on two different bases:
    - all costs and outcomes discounted at the prevailing rate recommended by the Treasury, currently 6% per annum
    - all costs and monetary outcomes discounted at the Treasury rate, currently 6%, but non-monetary outcomes not discounted.

    Both sets of results should be reported. The physical units and values of costs and outcomes prior to discounting should also be reported.
11. Sensitivity analysis should be conducted and reported. The sensitivity of results to all uncertainty in the study should be explored. This should involve the use of confidence intervals and/or ranges for key parameters, as appropriate. The ranges and choice of parameters to vary should be justified.
12. Comparisons with results from other studies should be handled with care. Particular attention should be paid to differences in methodology (such as the treatment of indirect cost) or differences in circumstances (such as different population groups).

Source: Association of the British Pharmaceutical Industry and the Department of Health (1994)

Chapter 5

Regulation, Competition and Equity: With or Without a Fixed Budget?

Wynand P.M.M. van de Ven

INTRODUCTION

Market-oriented health care reform is high on the political agenda in many countries (see e.g. Hurst, 1992; Saltman and von Otter, 1992; van de Ven, Schut and Rutten, 1994). Most of these reforms can be considered as the third stage in the development of arrangements for the financing and organization of health care (Hurst, 1990; Scotton, 1991). In the first stage the primary objective was the removal of (financial) barriers to access to health services. The second stage is characterized by the control of the subsequent rise in health care expenditures. The aim of the third stage is, within the constraints of equitable access and control of total health care expenditures, to improve the efficiency with which health services are produced and used.

Table 5.1 shows that during the 1960s and 1970s health care costs in Europe grew rapidly. This cost escalation is related to the economic growth, advances in medical technology, the extension of the health insurance coverage, and the relatively open-ended payment systems. Since the economic recession in the late 1970s/early 1980s the governments of many European countries focused on cost containment strategies in health care. These strategies appeared to be effective in decreasing the growth rates. Table 5.1 shows that in most European countries – in contrast to Canada and the United States – the increase in health expenditures during the 1980s has been quite modest. Abel-Smith (1992) surveyed the cost-control measures for health care in 12 European countries during the period 1983–1990. His major conclusion was that the European experience demonstrates that it is technically possible to control health care costs by government regulation of supply rather than demand, particularly by applying budgets to hospitals reinforced by manpower controls.

The success-indicator that Abel-Smith uses when he contends that regulation works in Europe, is cost containment. However, cost containment is not the same as

Fixing Health Budgets: Experience from Europe and North America.
Edited by F. W. Schwartz, H. Glennerster and R. B. Saltman. Technical editor R. Busse.
Published 1996 by John Wiley & Sons Ltd.

Table 5.1 Total health expenditures as a percentage of gross domestic product for 20 OECD-countries, 1960–1990

	1960	1970	1980	1990
Average of 18 European OECD-countries	3.7	5.1	7.0	7.5
Canada	5.5	7.1	7.4	9.3
United States	5.3	7.4	9.2	12.1

Source: Schieber, Poullier and Greenwald, 1992

efficiency. As Enthoven and Kronick (1989, p. 101) state: "None of the countries that have adopted fixed budgets in the public sector have solved the problem of creating incentives for the efficient organization and delivery of care." Indeed, based on a comparative analysis of seven European countries Hurst (1992, pp. 140–1) concluded that fixed budgets seem to be successful as cost containment strategy, but that micro-economic inefficiency emerged increasingly as the outstanding problem in the publicly financed health sectors of these countries. Micro-economic efficiency includes both static efficiency, that is improving value for money, and dynamic efficiency, that is, improvements over time in medical and organizational techniques which raise the productivity of scarce human and physical resources. Besides a lack of incentives for efficiency Hurst (1992, p. 142) mentioned the following other outstanding problems in several of these countries: excessive regulation, waiting lists, impersonal care, no responsiveness to consumer preferences, underservice, and inflexible and ineffective management systems. These problems are not costless. For example, two-year waiting for a new hip or six-month additional sick leave because of waiting for medical surgery is a real cost to society that does not show up in the health care costs statistics as presented in Table 5.1.

A crucial question is whether market-oriented health care reforms based on regulated competition can succeed in improving efficiency in health care within the constraints of equity and cost containment. By regulated competition I understand a competitive market in which the allocation and price setting are determined in principle by the market, but where government creates the necessary conditions ("rules of the game") in order to achieve society's goal with respect to equity and an efficient functioning of the market. Because we cannot exclude the option that a competitive health care system yields both more efficiency and higher total costs, politicians might be induced to put a fixed budget on health care expenditures. The term fixed budgeting refers to an overall budget limit on health care services, regardless of where the funds originate. This brings us to the question whether regulated competition in health care should (not) be combined with a fixed budget.

This chapter deals with this question, concentrating on the Dutch health care reforms. Launched in 1988, these are the most radical market-oriented health care reforms in Western Europe. The Dutch model can be considered the first attempt of a nationwide implementation of Enthoven's (1978) Consumer-Choice Health Plan, which by many economists is considered a prototype model that, at least in theory, closely approaches the optimal model for combining equity and efficiency in health care.

The following section outlines the Dutch reform proposal. The third section concentrates on equity, cost containment and efficiency in a regulated competitive health care market. Lastly, the conclusions will be summarized.

HEALTH CARE REFORMS IN THE NETHERLANDS

In 1988 the Dutch Government and Parliament decided to radically reform the health care system (see e.g. van de Ven, 1990). Although the implementation of the reforms is far behind schedule, some major steps toward a market-oriented health care system have already been realized (see e.g. van de Ven and Schut, 1994). The new government, which came into office in August 1994, will continue these market-oriented changes and intends to enforce an active anti-cartel policy in health care. Some elements of the original reform proposal, e.g. the transition path and the uniform regulatory regime for both catastrophic and non-catastrophic risks, will be changed.

The two key elements of the proposed system can be characterized as a compulsory health insurance for the whole population and regulated competition among insurers as well as among health care providers. The benefits package of the compulsory health insurance will be comprehensive. The insured will receive a subsidy to help them buy their compulsory health insurance. The subsidy will come from a Central Fund which will be filled with mandatory income-dependent contributions, to be paid to the tax-collector. From the Central Fund the subsidy will go directly to the qualified insurer chosen by the insured.[1] The subsidy per individual is independent of the chosen insurer and will be equal to the expected per capita costs within the risk group to which the insured belongs, minus a fixed amount, which is equal for all individuals. The deficit created by this deducted fixed amount is met by a flat rate premium to be paid by the insured directly to the insurer of his or her choice. The difference between the actual costs and the risk-adjusted payment will not be the same for all insurers and will be reflected in the flat rate premium that the competing insurers will quote. This creates the incentive for insurers to be efficient.

The insurers are expected to function as an intermediary between the consumer and the provider of care. To a high degree, insurers and providers will be free to negotiate conditions of contracts. In law the benefits package of the compulsory health insurance will be described not in terms of institutions such as hospitals or nursing homes, but as types of care. Any supplier meeting the quality standards may provide these services. This will greatly increase the possibilities for substitution of care. Insurers will be able selectively to contract with providers and to offer different insurance options as do health maintenance organizations, preferred provider organizations or traditional health insurers, as long as the insurance conditions provide the insured's coverage for all the types of care described in the law. Consumers will be free to choose their insurer, picking the policy they like the most. The premium paid will reflect the efficiency and cost-generating behaviour of the contracted health care providers.

EQUITY, COST CONTAINMENT AND EFFICIENCY

What can be Expected from Regulated Competition?

A major dilemma with respect to market-oriented health care reforms such as in the Netherlands is that politicians have expectations about the outcome of a competitive health care market that do not correspond with economic theory. Politicians strongly focus on cost containment in the sense of stabilizing the fraction of gross national product to be spent on health care. Politicians often consider regulated competition in health care as a substitute for detailed government regulation with respect to prices, volume and capacity in order to contain cost. However, economic theory predicts that under certain ideal conditions (e.g. open entry to the market, well-informed consumers, an adequate anti-cartel policy, transparency of the market and protection of the population against quackery), competition may improve the efficiency, quality, innovation and responsiveness to consumers' preferences. And none of that is synonymous with cost containment in the political sense. Therefore a crucial question is: can regulated competition improve efficiency in health care within the constraints of equitable access and control of total health care expenditures?

The condition of "equitable access" seems to be fulfilled in a regulated competitive health care system as proposed in the Netherlands. Via compulsory health insurance, each citizen has access to a broad benefits package of care with a high degree of solidarity between healthy and sick, between rich and poor, and between young and old. However, regulated competition cannot provide any guarantee for the fulfilment of the second condition, i.e. cost containment in the sense that the health care expenditures will not exceed a certain percentage of gross national product. Regulated competition can be expected to yield "more value for money". However, *if* additional health care can contribute to one's health and *if* competition in health care yields more value for money, and *if* most people judge that a good health status is the most important thing in life (more important than a good marriage, housing or job) – as does 60% of the Dutch population – then we can not exclude the option that a competitive health care system yields both more efficiency (lower unit costs) and higher total costs (units of higher quality and/or more units). For government such an outcome could be an argument to put a fixed budget (or cap) on the health care expenditures related to the compulsory health insurance, e.g. by putting an upper limit on the premium that each insured pays directly to the insurer.

Regulated Competition in Health Care: With or Without a Fixed Budget?

This brings us to the question of whether regulated competition in health care and a fixed budget can go hand in hand. In my opinion, there are two arguments to answer this question in the negative. First, assume that the fixed budget is exceeded and the cap becomes a binding constraint. Then government will intervene with price controls and controls on volume and capacity in health care, analogously to the situation during the last 20 years in the Netherlands and in many other countries. By definition this strongly reduces or even eliminates (the potential for) regulated competition. Second, if – as in the Netherlands – government wants to share the responsibility for cost containment and efficiency with market parties (insured, insurers, providers), isn't it then logical to let the market have a say in the optimal level and allocation of means to be spent on health care? What is wrong with higher health care costs if these are in accordance with the preferences of well-informed consumers? Nobody cares about the increased expenditures on cars, holidays and other luxuries during the last decades.

An important argument, however, for government to put an upper limit on some expenditures with respect to health care – even in a regulated competitive health care market – might be government's responsibility for providing *access* to health care for everyone. Access to care for sick and low-income people implies cross-subsidies from the healthy and high-income people. In the Dutch model these cross-subsidies are reflected in the income-related contributions to the Central Fund and the risk-adjusted subsidies from the Central Fund. If more care becomes available to the rich, then also the decent minimum that has to be available for everyone, probably will be set at a higher level. Therefore the subsidies and also the income-related contributions for the compulsory health insurance will go up. Ultimately this might have adverse effects on the economic process. This might be an argument for government to put an upper limit on the *public* expenditures on health care, i.e. the expenditures that go via the Central Fund.

Sometimes the *compulsory* character of health insurance is mentioned as an argument for a fixed budget on the *total* expenditures of the health insurance. In my opinion this argument is not convincing because without the obligation to buy health insurance, an overwhelming majority of the population would have bought it *voluntarily*. The compulsory character, however, does provide an argument to guarantee financial access to the compulsory health insurance for everyone and to carefully define the benefits package of the compulsory health insurance.

Consumer Choice of Insurance Contract

Given a cap on the expenditures that are necessary for the cross-subsidies to the sick and low-income people, a crucial question becomes: to what types of health care should the compulsory health insurance provide access? Should everybody have guaranteed access via the compulsory health insurance to *all* care with any positive benefits, independent of the costs? In 1991 the Dutch Government Committee on Choices in Health Care advised the government to exclude those benefits from the compulsory health insurance for which the costs are high and the expected benefits are very low.[2] These types of care will be referred to as "cost-ineffective" care. If the marginal value of the last dollar spent on health care is lower than the marginal value of the last dollar spent on goods other than health care, this type of rationing, which is based on the cost-effectiveness of health care, might improve overall efficiency in society.

According to the Dutch reform proposal the insured may choose among several insurance contracts offered by the insurers. One insurer may offer several insurance contracts as long as the insured's rights are in accordance with the law. The Government Committee on Choices in Health Care advised that insurers should be allowed to refuse to reimburse costs for inappropriate care, i.e. cost-ineffective care. The Committee called such an insurance contract the "Appropriate Care" policy. The insured would be free to buy, for example, a "Total Freedom" policy that unconditionally reimburses all health expenditures, but of course the premium of the latter policy would be much higher than that of the former policy. The risk-adjusted capitation payment from the Central Fund is only dependent on the health status of the insured, and not dependent on the chosen policy.

The difference between the two insurance contracts can be illustrated as follows. Suppose there is a certain disease or impairment that can be diagnosed and cured by different methods with different degrees of certainty and different prices:

method 1: 95% certainty, price $10

method 2: 99% certainty, price $100

method 3: 99.9% certainty, price $10 000.

The "Appropriate Care" policy with a relatively low premium would probably cover method 2 and not method 3, while the "Total Freedom" policy with a very high premium would cover method 3. Another example: suppose in a certain life-threatening situation there exists a treatment with a chance of 1% on success and with a price of $500 000. The "Appropriate Care" policy would not cover this treatment because of its low cost-effectiveness, while the "Total Freedom" policy might cover the treatment. The idea of competition between the "Appropriate Care" policy and the "Total Freedom" policy is closely related to Pauly's (1990) proposal for a model of consumer choice (with advice) among health plans, and competition among plans on the basis of the rate at which new technology is introduced.

Two-tier Health Care Acceptable?

The question is whether a two-tier health care system, in which those who can pay, will receive method 3 and the $500 000 treatment, and those who can *not* pay will *not* receive these treatments, is acceptable for European countries where solidarity and equity are always high on the political agenda. In answering this question one has to realize that *given* a limited amount of money available for the cross-subsidies from the healthy and high-income to the sick and low-income people, a two-tier system seems to be inevitable anyway. Those who can pay will always find their way, abroad or in their own country. For example, in the UK salaried physicians working in a hospital are allowed to treat private patients in their private practice for a private fee. Consequently, those who can afford can receive the treatment today, and those who cannot afford have to wait one or two years. The more successful cost containment is, the longer the waiting lists and the more the two-tiered system grows. In several Eastern European countries formally speaking health care is "free for all". However, in many cases patients have to pay "under the table payments" to the provider. So the total income of salaried physicians often is a multiple of their formal salary.

Given the inevitability of a two-tier health care I prefer a system with consumer choice among health plans in which everybody who needs it, has an immediate access to method 2, rather than a system without consumer choice and with waiting lists for methods 2 and 3. Those who think it is unfair that low-income people cannot afford method 3 and the $500 000 treatment, should realize that other health-influencing matters with probably a larger health effect, such as safe cars (with airbags and a side impact protection system) and good housing (no environmental pollution, no damp, and no breach of the peace), can freely be bought by those who can afford and are willing to pay.

CONCLUSION

In this chapter I have dealt with the question whether in a regulated competitive health care market government should put a fixed budget on health care expenditures. As a case study I have concentrated on the (proposed) Dutch health care reforms. The primary task of government with respect to health care is *not* to contain costs, but to guarantee financial access to care for everyone ("equity") and – in case of a regulated competitive health care market – to let the market function well ("efficiency"). If the conditions for a well functioning of the health care market are fulfilled, it seems logical from an efficiency point of view to let the market have a say in the optimal level and allocation of health care spending.[3] So then Government should *not* put a fixed budget on *total* health care expenditures. However, if as a result of competition total health care expenditures go up, the cross-subsidies which are necessary to guarantee access to care for the sick and low-income people may also go up. This might ultimately have an adverse effect on the economic process and thereby induce politicians to put an upper limit on the *public* expenditures, i.e. the expenditures which are necessary for the cross-subsidies to the sick and low-income people. The Dutch Government Committee on Choices in Health Care advised that in such a situation one should exclude those benefits from the compulsory health insurance for which the costs are high and the expected benefits are very low. This new concept of "equity", i.e. "equal access to appropriate or cost-effective care", induces a two-tier health care system in which only those who are able to pay, have access to care outside the compulsory benefits package. However, given a limited amount of money which is available for the cross-subsidies to the sick and low-income people, a two-tier health care system is inevitable anyway. The cost-effectiveness of care then seems to be a defensible criterion on the basis of which society decides to which types of care everyone should have access.

ACKNOWLEDGEMENTS

I would like to thank Erik Schut for his valuable comments on a previous version of this chapter. I am, of course, responsible for its content.

This chapter is reprinted with permission from: W.P.M.M. van de Ven (1995), Regulated competition in health care: with or without a global budget? *European Economic Review*, **39**, 3–4, 786–94, with permission from Elsevier Science BV, Amsterdam, The Netherlands.

NOTES

1. The new Government that came into office in August 1994 proposed a different system of cross-subsidies for the higher income groups and the self-employed. In this chapter I will further concentrate on the payment system as proposed for the majority of the population. *Mutatis mutandis* the same arguments hold for the others.
2. *Choices in Health Care*, November 1991. A report by the Government Committee on Choices in Health Care, The Netherlands. Distributed by: Ministry of Welfare, Health and Cultural Affairs, PO Box 5406, 2280 HK Rijswijk, the Netherlands.
3. According as these conditions are not fulfilled and market imperfections exist, Government may feel itself responsible for the level of health care expenditures.

REFERENCES

Abel-Smith, B. (1992). Cost containment and new priorities in the European Community. *The Milbank Quarterly*, **70**, 393–416.

Enthoven, A.C. (1978). Consumer-choice health plan; a national-health-insurance proposal based on regulated competition in the private sector. *New England Journal of Medicine*, **198**, 650–8 and 709–20.

Enthoven, A.C. and Kronick, R. (1989). A consumer-choice health plan for the 1990s. *New England Journal of Medicine*, **320**, 29–37 and 94–101.

Hurst, J. (1990). The reform of health care systems in seven OECD countries. Paper presented at the Second World Congress on Health Economics, Zürich.

Hurst, J. (1992). *The Reform of Health Care; A Comparative Analysis of Seven OECD Countries*. Paris: OECD.

Pauly, M.V. (1990). Financing health care. *Quarterly Review of Economics and Business*, **30**, 63–80.

Saltman R.B. and von Otter, C. (1992). Planned markets and public competition. Buckingham: Open University Press.

Schieber, G.J., Poullier, J.P. and Greenwald, L.M. (1992). US health expenditures performance: an international comparison and data update. *Health Care Financing Review*, **13**, 4, 1–88, Appendix, Tables 1 and 33.

Scotton, R.B. (1991). National health insurance in Australia: new concepts and new applications. National Centre for Health Program Evaluation, Working Paper no. 11, Melbourne.

van de Ven, W.P.M.M. (1990). From regulated cartel to regulated competition in the Dutch health care system. *European Economic Review*, **34**, 632–45.

van de Ven, W.P.M.M. and Schut, F.T. (1994). Should catastrophic risk be included in a regulated competitive health insurance market? *Social Science and Medicine*, **39**, 1459–72.

van de Ven, W.P.M.M., Schut, F.T. and Rutten, F.F.H. (1994). Forming and reforming the market for third-party purchasing of health care. *Social Science and Medicine*, **39**, 1405–12.

Part 2

Sectoral and Regional Dilemmas

Chapter 6

Fixed Budgets in the Inpatient Sector: the Case of the Netherlands

Hans A.M. Maarse

INTRODUCTION

Cost containment has been a top issue in Dutch health care policy making since the late 1970s, although concerns about the escalating growth of health care expenditure date from early in that decade, when the annual growth rate varied between 12% and 16%. In particular after the first world oil crisis in 1974, policy makers became increasingly worried about the disproportional rise of health care expenditure compared with the growth of the national economy. The share of health care expenditure in gross domestic product rose by 2.2% from 6.1% in 1970 to 8.3% in 1980. It was generally felt, that the rapid growth of health care expenditure had to be stopped. Otherwise, health care would consume too many resources.

The need for cost containment has led to the introduction of many new policy tools. From the very beginning a strong emphasis has been put upon the control of the volume of health care facilities (e.g. hospital beds, medical specialists). In fact, "action on supply" has always played a much stronger role in Dutch health care cost containment than "action on demand" (the terminology is from Abel-Smith, 1992). In the 1980s, the repertory of supply-side instruments was extended by the adoption of fixed budgets for all providers of health services. At present, fixed budgets can certainly be considered a cornerstone of health care cost containment in the Netherlands.

This chapter consists of two parts. The first part addresses some general aspects of fixed budgets in Dutch health care policy. The second part focuses upon fixed budgets for hospital care. It is important to note that no attention will be given to the issue of internal budgeting within provider institutions. The use of fixed budgets to control the health care expenditures of the sickness funds will also be left out of consideration (see Chapter 5 by van de Ven).

Fixing Health Budgets: Experience from Europe and North America.
Edited by F. W. Schwartz, H. Glennerster and R. B. Saltman. Technical editor R. Busse.

GENERAL ASPECTS OF FIXED BUDGETS

The use of fixed budgets stems from the early days of cost containment. In 1974 the Minister of Health put an annual budget limit to the total volume of capital investments in hospital construction projects. The limits were reasonably effective: over the period 1974–82 the total volume of investments did not exceed the annual limit, or did so only marginally. The financial success was accompanied, however, by the creation of a large reservoir of urgent hospital investment projects for which no authorization had been given.

Fixed budgets rapidly gained popularity among health care policy makers in the 1980s. In 1983 the Minister of Health imposed a fixed budget upon the acute hospital sector. Other health care facilities, such as nursing homes or psychiatric hospitals, followed shortly afterwards. Health care delivery by medical specialists has been subjected to a fixed budget since 1989. Interestingly, this budget was not an initiative of the government but the result of self-regulatory action of the representative associations of the medical specialists, health insurers and hospitals. They negotiated a package deal on the remuneration of medical specialists, which included the introduction of a "macro-budget" in order to limit the expenses for specialist care (Lieverdink and Maarse, 1995). At present, fixed budgets are common for all providers of health care. The Financial Report on Health (*Financieel Overzicht Zorg*), which is published annually by the Minister of Health, gives an overview of the fixed budgets for the following year for each health care sector. It also evaluates health care expenditures in recent years. In case of budget overruns, compensatory measures are announced.

Policy makers consider fixed budgets an attractive cost containment instrument. They are dissatisfied with the results of many other instruments. Action on demand by the introduction of user charges has always met much political resistance in the Netherlands and their financial results are questionable. Bed-reduction programs have been only partly effective and their direct financial impact is unclear. Policy makers also strongly believe that health care cost containment will be doomed to fail, if funding arrangements remain open-ended. Fixed budgets are necessary to limit total expenditure for, say, acute hospital care or dental care.

Expenditure Targets versus Expenditure Caps

The terminology of a fixed budget is somewhat misleading in the Dutch context, as it suggests that a fixed amount of financial resources is made available to the providers of health care. This is only true for the tax-funded part of Dutch health care, which is relatively small (10.3% in 1991). In order to clarify the concept, Glaser's distinction between an expenditure cap and an expenditure target is helpful. He defines an *expenditure cap* as: "a fixed amount of money. All services delivered during the year are contained within that sum." An *expenditure target* is quite a different thing. It is a "goal for the current and, possibly, subsequent years" (Glaser, 1993, p. 98).

The concept of a fixed budget in Dutch health care policy making comes much closer to what Glaser calls an expenditure target than to an expenditure cap. According to Glaser, "caps imply strong control, usually within government authority. All money passes through either a centralized payer or several highly coordinated payers" (p. 98). The condition of strong control is only partly met in Dutch health care, however, not only because health care is financed through a multiple payer system (government, sickness funds, private health insurers, patients), but also because in some health care sectors (e.g. hospital care) decision making on fixed budgets is to a considerable extent decentralized to the health insurers. Another, perhaps more compelling, reason why the concept of an expenditure target should be preferred stems from the prominent role of fee-for-service funding in Dutch health care. In Holland fixed budgets usually do not mean that a fixed amount of resources is channeled from a payer or a group of payers to a provider. A fixed budget only limits the total allowable amount of resources to be spent by a provider or group of providers in a fiscal year. The provider remains to be paid on a fee-for-service basis. If the provider exceeds the budget limit, the instrument of fee cuts in subsequent years is used to compensate for the budget overrun.

Sector Budgets and Provider Budgets

It is important to make a distinction between two different types of fixed budgets, namely sector budgets and provider budgets. The distinction can be easily illustrated for hospital care. Whereas a *sector budget* puts a limit to total expenditure for all hospital care, a *provider budget* puts a limit to total expenditure of a single hospital. Sector budgets have become common across health care. There are sector budgets now for hospital care, psychiatric hospital care, ambulatory psychiatric care, nursing home care, care for the mentally handicapped, dental care, medical specialist care, physiotherapy, general practitioner care, prescription drugs, and so on. Provider budgets are in principle only used for provider institutions, e.g. hospitals, nursing homes or centers for ambulatory psychiatric care. They are not common for individual providers, e.g. general practitioners, medical specialists or dentists (physiotherapists present an exception). Physicians are only imposed a sector budget, not a provider budget.

The distinction between sector and provider budgets is crucial. It raises the question who decides upon sector budgets and provider budgets. In the Netherlands, the Minister of Health decides on sector budgets after the government has determined the total amount of financial resources available for health care. This is done in the Financial Report on Health. Procedures and the extent of government involvement in decision making on provider budgets differ per sector. The health insurers must negotiate production contracts with the provider institutions.

The distinction between sector and provider budgets also demands an allocation model in order to apportion a sector budget into separate provider budgets. Allocation models are often very complicated and subject to frequent change. A standard issue is to what extent allocation must follow historical patterns or must be based upon other criteria.

A final aspect of the distinction points to budget overspendings. If a single provider overspends its budget, it is fair (at least in principle) to impose upon this provider a budget cut to compensate for the overspending. However, if the total sum of provider budgets exceeds the sector budget, for instance because the total amount of production contracts between the health insurers and hospitals surmounts the budget limit made available by the Minister of Health, a mechanism must be found to offset collectively the overspending. A collective compensation mechanism is also needed, if the sector budget is overspent and there are no provider budgets.

Public versus Private Expenditure

Until recently, most fixed budgets included both public and private health care expenditures. This comprehensive approach follows from the Dutch idea that cost containment programs must control the total volume of health care expenditure, not only the volume of public expenditure. Thus, the fixed budget for hospital care comprises both public and private expenditures for hospital care.

The fact that the distinction between public and private health expenditures has always been considered irrelevant in most cost containment programs and fixed budgets setting deserves some explanation, because it is, at least from an international point of view, a rather exceptional strategy (Abel-Smith, 1992). It cannot be seen apart from the traditionally prominent role of private (voluntary) health insurance in Dutch health care finance. This prominent role is directly related to an arrangement in the Sickness Fund Act which rules that a person whose income exceeds the government-set income ceiling for the sickness fund scheme must automatically leave the sickness fund and buy a private health insurance plan. An opt-out provision as exists in Germany is not provided in the law (Maarse, 1995a; Schut, 1995). In 1993 about 33% of the population had private health insurance. The share of private health insurance in health care finance was 14.1%.

The comprehensive approach to cost containment and fixed budgets rests upon two arguments. First, it is argued that private health expenditures have in fact a similar effect on employment, labour costs and competitiveness as public expenditures, since widespread risk aversion means that almost nobody is really free to decide whether or not to purchase private health insurance. Secondly, policy makers often argue that a more limited approach to cost containment, focusing only on the public part of health expenditures, would eventually result in a two-layer health care system with minimum facilities for the lower layer and generous facilities for the upper layer. Such a system would violate the principle of equity which is traditionally highly valued in Dutch health care.

There are some exceptions to this comprehensive approach. In physiotherapy, for example, the fixed budget applies only to sickness fund patients. If a physiotherapist has "consumed" his or her budget for sickness fund patients (which is defined as a maximum number of treatments), he or she is still free to treat patients with private health insurance. This arrangement explains why sickness fund patients may face waiting lists for physiotherapy, whereas patients with private health insurance do not. A similar situation occurs in orthodontics.

Interestingly, the comprehensive approach to cost containment and fixed budgets seems to be showing the first signs of erosion. In 1995, the government has decided to implement a considerable reduction in dental care benefits for sickness fund patients. They may voluntarily opt for private supplementary health insurance to cover (part of) their expenses for dental care. After the reduction, the fixed budget will only apply to those benefits which are covered by the sickness fund scheme and similar benefits under private health insurance. All other expenditures for dental care are regarded beyond the control and responsibility of the Minister of Health.

Decision Making on Fixed Budgets

The introduction of fixed budgets raises the problem how to decide on how many resources should be spent on, for instance, hospital care, psychiatric care or dental care. What should the budget be of a single hospital, a nursing home or a center for ambulatory psychiatric care?

There are two different approaches to decide budgets. In the *rational* approach budget decisions are primarily guided by information, for instance on the (changing) demand for health care, on the impact of technological innovation or on costs. As regards provider budgets, deliberate allocation models are constructed which must establish a fair link between budgets and the demand for health services. Allocation models are made up of budget rules which must be strictly applied. In the *political* approach budget decisions are mainly the result of a negotiating process between agencies. Decisions are shaped by lack of information, by a selective use of information, by conflicts of interests among the participants, by power relations and by the permanent need to compromise.

Decision making on sector budgets is a mixture of the rational and political approach. But it seems plausible to assume that decision making is more shaped by political forces than by information. Of course, health care policy makers will always argue that their budget decisions were guided by information and that their budget decisions reasonably match the growing demand for health care. The difficulty, however, is that they often lack necessary information. Much is unknown, for instance, about how demographic, social, cultural, epidemiological and other factors affect health care demand and, subsequently, health care expenditures. Information on the effects of technological innovation and on productivity gains in health care delivery is also poor. Therefore, decisions must often be based upon assumptions that are shaped by political considerations.

Information is also often conflicting. The Minister of Health estimated the growth in the demand for health care over the period 1994–98 at 2.3% a year. But another commission which worked for the Dutch Care Federation (Nederlandse Zorgfederatie) estimated a 2.6 percentage (van der Zwan, 1995). Which figure is accepted seems to be a matter of political preference.

Even when policy makers were perfectly informed about health care demand, political decision making would still play a prominent role. Fiscal stringency demands hard political decisions on expenditure cuts in health care. Decisions on sector budgets are increasingly dictated by the need to stay within the overall budget for health care. The need for political compromises at the government level is nicely demonstrated by the fact that in spite of the estimated annual growth rate of 2.3%, the allowable growth rate was fixed at a 1.3% level (*Financieel Overzicht Zorg, 1995*).

Health care policy makers have also spent much energy on rationalizing decisions on provider budgets through the development of rational allocation models. These models are often very complicated and consist of pages of detailed budget rules, written by the Central Agency for Health Care Tariffs (Centraal Orgaan Tarieven Gezondheidszorg) which can best be understood as a quasi non-governmental body (Lieverdink and Maarse, 1995).

Unfortunately, there are no studies available on how health policy makers decide allocation models. How should the model look? Which criteria should be used to determine a provider budget and which weights should be assigned to each of them? Who should participate in the decision-making process? Policy makers will seek for budget rules which give a good fit between budget and expenditures. This is the rational approach to model building and rule-making.

But decisions on allocation models also have a political dimension. Take for instance the case of hospitals. The interests of big hospitals may conflict with those of small hospitals. The interests of hospitals in urban areas may conflict with those of hospitals in rural areas. Changes in allocation models may cause big budget shifts between hospitals which are heavily opposed by the "budget losers." The participants in the decision-making process will attempt to use their influence to shape the final outcome in their own interest.

We shall illustrate the politics of model building with two other examples. In the new allocation model for hospital care, hospitals are paid an earmarked budget for the treatment of AIDS patients. From the rational perspective, such a budget can be justified with the argument that the treatment of AIDS patients is very costly. From the political perspective, the question must be raised: why a separate budget for treating AIDS patients and not for any other patient group whose treatment is also very expensive? The answer would be that the AIDS lobby has been more successful than any other patient interest group in advocating an earmarked budget. The second example points to the decision not to include the revenues of the medical specialists in the hospital budget. This decision was a political compromise to overcome the specialists' objections against fixed hospital budgets.

Putting allocation models into practice is also a mixture of rational and political elements. Rules must be strictly applied (rational approach) but, as the implementation literature (e.g. Berman, 1978; Mayntz, 1984) has spelt out, even strict rules may leave much room for interpretation and negotiating between the implementing agencies. Rules and procedures often appear multi-interpretable, when it comes to implementation. Which interpretation is chosen, may be a political decision. Take again the case of hospital budgets. As we will see in the next section, hospital budgets are partly dependent upon the number of hospital beds. Hospitals receive a fixed amount of money for each bed they have. But how many beds does a hospital have? The answer on this simple question may be complicated, for instance in the situation in which the hospital must reduce its number of beds. Which number of beds must be used in calculating the hospital budget: the number of beds before or after reduction? This question may turn into an issue for negotiation between the hospital and the Minister of Health. The situation may even become more complicated, when the hospital challenges the imposed reduction in court.

Another aspect of the political dimension in decision making on provider budgets points to the prominent role of production contracts which must be negotiated between the providers and the health insurers. These negotiations have been described as mixed-motive games. Whereas the providers have an interest in generous contracts, the insurers do not. At the same time both parties have a common interest in finding a compromise on a production contract because of their common responsibility for the delivery of good health care (Maarse, van der Horst and Molin, 1993).

The Implementation of Fixed Budgets

After decisions on fixed sector and provider budgets have been taken, the question arises how to ensure that the budget is not overspent. This is the implementation dimension of fixed budgets.

The implementation of fixed budgets is a difficult enterprise in the Netherlands. Lack of information presents a big problem. In order to find out whether expenditures have stayed within the predetermined budget limits, timely and reliable information is necessary. This requirement is often not fulfilled. In spite of a profusion of statistical data, the quality of information on health care expenditures is still poor. The Ministry of Health is dependent on information which comes from various sources. Between these sources there is much variation in the definition of concepts and methods for data collection. Figures on sectoral budgets and expenditures may be only partly comparable, because figures on expenditures include costs which are not included in the sector budget and vice versa. Attempts to improve information on health care expenditures are now underway, but there is still a long way to go. The ongoing integration of the statistical data of the sickness funds and private health insurers can be viewed as a big step forwards. Another source of confusion relates to the fact that budgets are often due to retrospective changes.

Information problems do not only cause confusion, but also raise conflicts between the Minister of Health and the providers' associations on the size of budget overruns. It often takes much negotiation to achieve consensus on expenditure data (Lieverdink and Maarse, 1995). The dependency of the Minister of Health on the suppliers of information in data collecting may prompt a strategic delivery of information by the providers.

A second implementation problem relates to the role of production contracts in decision making on provider budgets. There is no arrangement in the allocation model to control the total volume of production contracts. These contracts often appear a source of overspending.

A third implementation problem is that fixed budgets do not directly regulate the volume of medical care. They fix only the total budget for medical care. Difficulties particularly arise in those sectors where no provider budgets exist and where physicians are paid on a fee-for-service basis. An intermediary structure between the providers and health insurers such as the Kassenärtzliche Vereinigungen in Germany, which monitors the volume of medical services and adjusts the fees paid to the physicians over the year in order to avoid overspendings, does not exist in the Netherlands. If it turns out at the end of the year or even later, that spendings have not stayed within the fixed budget, the Minister of Health must use the instrument of fee-cuts in *subsequent* years to offset the overspending. Fee-cuts may be either uniform for all physicians or selective, that is targeted at specific groups of physicians. The last procedure is used when certain specialties can be identified which are held responsible for the overspendings.

The absence of an institution which monitors the volume of medical care and adjusts fees during the year for which the fixed budget was set, presents a serious structural weakness with even counterproductive effects. It may be hypothesized that retrospective fee-cuts will boost the volume of medical services, since physicians are facing a prisoner's dilemma. Since the medical production of his fellow physicians affects his own income, each physician will feel a strong incentive to increase his medical production to counteract a loss of income because of lower fees. This can be achieved by doing more or by charging differently. There is some empirical confirmation for this hypothesis. The average annual growth in medical services sharply increased after the introduction of a fixed sector budget for specialist care in 1989: the volume of inpatient treatments rose from 1.0% over the period 1982–88 to 2.1% over the period 1989–92 and the volume of outpatient treatments from 2.4% to 4.2% (*Financieel Overzicht Zorg 1994*). Obviously, retrospective cuts in fees also create much frustration among physicians who perceive the dynamics of fixed budgets as "doing more and getting less."

This brief discussion of implementation problems demonstrates that the notion of the implementation of fixed budgets as a purely administrative game is wrong. Implementation is also a political process, in particular when it comes to decision making on the size of overspendings and the fee-cuts which are necessary to compensate for them.

Are Fixed Budgets Effective?

The financial effectiveness of fixed budgets refers to the extent that expenditures are really kept within the predetermined limits. The measurement of effectiveness is often difficult. One source of problems is that fixed budgets are subject to frequent changes, creating confusion about what the budget really looks like. It should also be recalled here that information on budgets and expenditure may not be entirely comparable, due to differences in definitions. Finally, information problems on expenditures may arise among the participants in the budget game. This problem was discussed earlier.

As for the empirical results, it is important to distinguish between sector budgets *with* provider budgets (e.g. nursing home care) and sector budgets *without* provider budgets (most physician care). Let us first briefly discuss the results of sector budgets with provider budgets. The results for acute hospitals will be considered in the next section on hospital budgeting.

Evaluations in the *Financial Report on Health 1995* show that fixed budgets in 1992 and 1993 for psychiatric hospitals, centers for ambulatory psychiatric care, nursing homes and for institutions which provide care to the mentally handicapped, have been reasonably effective at the aggregate level. Overspending was only marginal – mostly less than 0.5% – and there was also marginal underspending. Production contracts often caused a (small) overspending, which was often partly offset, however, through delays in building activities causing underspending. Other incidental factors may also explain why expenditures deviate from budgets (e.g. statistical corrections). Compensation strategies depend on how the factors explaining over and underspending are assessed. Overspendings which are defined as structural must be compensated through expenditure cuts in subsequent years.

Sector budgets without provider budgets were less effective. Over the period 1980–89 aggregate expenditure for specialist care grew by an average of 2.6% a year. This average rose to 6.3% over the period 1990–92, whereas the specialists had accepted a zero-growth. The large overspending led to much political trouble. In 1991 long negotiations between the Minister of Health and the Dutch Association of the Medical Specialists (Landelijke Specialistenvereniging) were needed to find an agreement on a general fee-cut. In 1992 and 1993 the Minister imposed two unilateral cuts on the specialists' fees, because an agreement on cuts with the LSV could not be achieved. A large overspending was also reported for self-employed psychiatrists (Dfl 151 million in 1993), for general practitioners (+2% in 1993) and for dentists (+6.1% in 1993). The tool of retrospective fee-cuts was used to compensate for all these overspendings.

FIXED HOSPITAL BUDGETS

Until 1983, the funding arrangement for hospital care was open-ended. Dozens of guidelines of the Central Agency for Hospital Tariffs regulated the allowable costs for hospital care. The guidelines linked the allowable costs for personnel to the expected volume of medical care (inpatient days, occupancy rate); they did not control the volume of medical care itself. The budget estimate was screened by the local sickness fund. It was divided by the expected number of inpatient days in order to calculate the patient per diem rates which were used as the main method of payment. Furthermore, hospitals charged the health insurers for the costs of specific services (e.g. outpatient care) which were not included in the per diem rate.

This funding arrangement was a source of growing concern among health care policy makers who argued that it was entirely inconsistent with the goal of cost containment. It gave hospitals a strong financial interest in a high production of hospital services. It did not stimulate them to control costs and to increase efficiency. Another handicap was the labyrinth of guidelines which limited the autonomy of hospitals.

The introduction of fixed budgets in 1983 meant that the Minister of Health imposed a sector budget for hospital care (one for general and special hospitals and one for academic hospitals). The sector budget was fixed at the 1982 level plus some adjustments. Next, a prospective budget (the allowable costs) was imposed upon each hospital which covered most of its expenses. Hospitals could no longer influence their budget by doing more. The fixed provider budget meant that the traditional link between the volume of hospital services and revenues had been cut. Because the new funding arrangement had to improve efficiency, hospitals could add the full surplus to their reserves. But they were also made responsible for their deficits. Budget adjustments to relieve financial problems were abolished (Maarse, 1995b).

The hospital budget did not cover the revenues of the medical specialists who work on a fee-for-service basis in the hospital. The LSV successfully opposed the proposal to integrate the specialists' revenues into the hospital budget. The physicians perceived such an arrangement as an unacceptable restriction to their professional autonomy and feared its implications for their personal revenues (Lieverdink and Maarse, 1995).

The Allocation Model: from Historical to Functional Budgets

When budgeting was introduced in 1983, an allocation model for deciding provider (hospital) budgets was lacking. Hence, a pragmatic approach was adopted. Each hospital received a budget equal to its 1982 level of expenses plus an adjustment for inflation and some other cost increases. This was the model of *historical* budgeting. An important advantage of this model was that it made a rapid adoption of hospital budgeting feasible and did not cause major budget-shifts between hospitals. But it also led to problems. The 1982 level of expenses had been accepted as the norm. Hospitals with a relatively low level of expenses in 1982 repeatedly claimed, that historical budgeting in fact punished efficient hospitals and rewarded inefficient hospitals. Another problem was its inflexibility. Changes in the hospital workload were not taken into account. Adjusting budgets to changes in hospital capacity (beds, medical specialists) also proved a difficult exercise in the absence of clear guidelines.

These problems led in 1988 to the introduction of the model of *functional* budgeting which is still operative. The essence of the functional allocation model is that hospitals should be paid an equal budget for performing equal tasks. The model distinguishes three budget components: an availability component, a capacity component and a production component. The *availability* component is measured as the size of the clinical catchment area of the hospital and the *capacity* component as the number of authorized beds and medical specialist units which render outpatient care. The *production* component requires annual negotiations between hospital management and the health insurers about the volume of hospital care. Production contracts have to be made upon the following services: number of hospital admissions, inpatient days, first outpatient visits and day care visits. Additional contracts are required for some specific high-cost treatments such as cardiac surgery or renal dialysis. In 1992, the availability component accounted for 15% of the aggregate budget for hospital care, the capacity component for 34% and the production component for 51% (*Financieel Overzicht Zorg 1992*). The availability and capacity component are assumed to cover the fixed part of hospital expenses and the production component the variable part.

In order to calculate the hospital budget, the scores of the hospital on each of the three components must be multiplied with the rates which are set by the COTG. In 1992, for instance, the rate for a hospital bed was Dfl 11 000, for a specialist unit Dfl 353 000 (on average), for a hospital admission Dfl 1150, for an inpatient day Dfl 60 and for a first outpatient visit Dfl150, and so on. The rates for hospital services vary with the size of the hospitals, big hospitals receiving a higher rate than a small hospitals because they are assumed to perform more complicated and thus more expensive treatments. Hospitals are also paid additional budgets for interest payments, investments, salaried physicians, and a number of other items.

The transition from historical to functional budgets caused major budget shifts between hospitals. The difference between the hospital's historical and functional budget is called a reallocation effect. This effect can either be positive, indicating that the hospital was underfunded under the historical allocation model, or negative, indicating that it was overfunded. Reallocation effects may be substantial: whereas some hospitals faced a negative reallocation of more than 8%, others faced a positive reallocation of more than 8%. In order to dampen these effects, functional budgeting was phased in step-wise.

The inclusion of a capacity component into the allocation model means that the Minister of Health is indirectly involved in decision making on hospital budgets, because each hospital needs an authorization of all beds and specialist units to get paid for them by the sickness funds. An implication of this arrangement is that (delays in) bed-reduction programs have immediate budgetary consequences. The relative high percentage of production contracts points to a substantial degree of decentralization in the budgeting process. It has strengthened the role of the sickness funds and private health insurers which was rather marginal under the historical allocation model. Production contracts are seen as a tool to adapt budgets to changes in demand for hospital care. They must make decision making on provider budgets more flexible.

In 1992 the model of functional budgeting was revised. Some important changes were: (a) raising the rate of day care visits in order to stimulate substitution of patient care by day care; (b) weighting the rates for admissions and first outpatient visits; (c) lowering the share of the availability component in the hospital budget in favour of the share of the production component (51%) so greater emphasis is on production-bound factors and the flexibility of hospital budgeting is increased; (d) adding radiotherapy to the list of specific high-cost treatments for which separate production contracts must be made; (e) halvation of the so-called merger effect (see below).

Payment

The determination of a hospital budget must be kept apart from the payment of this budget by the health insurers to the hospital. Inpatient per diem rates have remained the most important payment method. The rate is calculated as follows. The first step is to calculate the total allowable costs, that is the fixed hospital budget. Next, the expected revenues of specific services for which the hospital charges the health insurers separately, are subtracted from the budget. Finally, the remaining sum is divided by the negotiated number of inpatient days. In 1992 patient per diem rates accounted for about 85% of total hospital revenues.

A revision of the payment method was initiated in 1993. The goal of this revision was to achieve all-outpatient per diem rates through a considerable extension of the list of services for which the hospital charges the health insurers separately (output- pricing). Hospitals must charge health insurers separately for all surgery, all more expensive diagnostics and treatments of patients, outpatient visits and day care. Obviously, this revision will cause a drop of patient per diem charges which are now expected to account for about 70% of total hospital revenues.

The introduction of the functional allocation model led to a sharp rise of the patient per diem rates. This is partly due to the fact that the introduction of fixed hospital budgets accelerated the drop in the volume of patient days. Hospitals see the reduction of inpatient days by reducing the length of stay and substituting inpatient care through outpatient care as an effective way to make savings on personnel costs (Maarse, van der Horst and Molin, 1993). Another cause for the sharp rise of rates is that in many hospitals the actual volume of patient days fell short of the negotiated volume. Since the budget is guaranteed, surcharges in subsequent years were needed to compensate the hospital for the loss of revenues due to underproduction. The paradox is thus that a more efficient delivery of health services, resulting in a *decrease* of inpatient days through a substitution of inpatient care by outpatient and day care, leads to an *increase* of patient per diem rates. Delays in payment due to underproduction have been identified a serious policy problem, the more so because hospitals must cover their deficits with private loans for which they must pay interest. These payments are subject to full reimbursement. According to COTG estimates, total delays amounted to about Dfl 1.7 billion at the end of 1991. Initiatives to remove them have been successful.

Capital Investments

According to the Hospital Facilities Act (*Wet Ziekenhuisvoorzieningen*) investments in construction activities require government approval. Hospitals must contract loans from private banks in order to finance their investments. Depreciations and interest payments are fully incorporated in the inpatient per diem charges. Contrary to, for instance, Germany, there is no dual financing system in the Netherlands: both operating and capital expenses are paid by health insurance. Until recently, hospital loans were guaranteed by the government, which made the hospital sector an interesting partner for the banking sector. This arrangement has now been ended in order to encourage hospitals to operate as a market agency in negotiating loans for their capital investments. The banking sector regretted this change and even alleged that hospitals have lost much of their former attractiveness as partners for investments.

Article 18 of the Hospital Facilities Act stipulates that hospitals must also have a government authorization for delivering certain types of top-clinical care. The rationale of this Article is to concentrate facilities, to improve the quality of care, to reduce costs and to prevent uncontrolled growth. Facilities can only be bought under Article 18, if they are very expensive, require much specialized expertise or if they are controversial from an ethical point of view. Article 18 applies for the following facilities: renal dialysis, kidney transplantation, radiotherapy, complex neurosurgery, heart surgery, PTCA, neonatology, clinical genetic research, IVF, bone marrow transplantation, liver and pancreas transplantation and positon emission tomography.

The investments for these forms of top-clinical care are covered by inpatient per diem charges and by special tariffs health insurers must pay if an insured person has received the treatment. The health insurers must also negotiate production contracts with the hospitals on maximum number of treatments.

Financial Effects

Table 6.1 compares the sector budget for hospital care with hospital expenditures over the period 1978–92. The figures in columns A and B indicate that the growth in the sector budget for hospital care has been very limited during the first years of hospital budgeting. The figures confirm that fixed budgets have been used only to impose large-scale expenditure cuts upon the hospital sector. This pattern has considerably changed since 1988, however. The major reason for raising the sector budget in the 1990s was not to meet the growing demand for hospital care but inflation and wage-increases. This observation is in fact alarming from the viewpoint of cost control, because it suggests that budgetary restraint is only a temporary option. After some period of time hospital expenses will inevitably increase again (Schwartz, 1987). Adjustments of the sector budget in order to decrease the work pressure of the nursing staff point into the same direction. These adjustments were the result of political action of hospital workers who no longer accepted the consequences of budgetary restraint.

Columns C and D suggest three different periods in hospital expenses. During the pre-budgeting period (1978–82) the annual growth rate of hospital expenses was 8.7%. This rate was radically slowed down after the introduction of hospital budgeting (1.3% over the period 1983–88). Since 1989 hospital expenses rise again. The annual growth rate over the period 1989–92 is 7.7%.

Sector budgets and hospital expenses are compared in column E. The figures demonstrate that hospital budgeting has been rather successful over the period 1983–88. The positive result in 1983 was largely due to a budget adjustment in late 1983 which did not lead to an increase of expenditures in 1983. Over the period 1989–91, however, sizeable gaps between expenditure targets and hospital expenditures can be observed. Quite remarkably, they coincide with considerable increases in the sector budgets (see columns A and B).

There are a number of complementary explanations for the gap between sector budgets

Table 6.1 Hospital budgets and hospital spendings: 1979–1992

	Sector budget for hospital care	Percentage change in sector budget	Total hospital expenses	Percentage change in total expenses	Difference between budget and expenses
1979	–	–	6 995	9.0	–
1980	–	–	7 639	9.2	–
1981	–	–	8 219	7.6	–
1982	–	–	8 812	7.2	–
1983	8 812 (HB)	–	8 874	0.7	0.1
1984	9 058	2.0	8 835	–0.4	2.5
1985	9 097	0.4	9 043	2.3	0.6
1986	9 165	0.7	9 228	2.0	–0.7
1987	9 242	0.8	9 250	0.2	0.1
1988	9 457 (FB)	2.3	9 439	2.0	0.2
1989	9 706	2.6	9 868	4.5	–1.7
1990	10 434	7.5	10 660	8.0	–2.2
1991	11 163	7.0	11 490	7.8	–2.9
1992	11 778	5.5	12 148	5.7	–3.1

Note: HB = historical budgets, FB = functional budgets
Source: Ministry of Health

and expenditures. The first explanation is that the demand for hospital care has never been taken into account sufficiently in government decision making on sector budgets. Decisions on sector budgets were primarily political decisions, dictated by the necessity to constrain spendings. They did not reflect the growth in workload of hospitals. Similarly, cost increases due to technological innovations (for instance new pharmaceuticals) may be neglected in the decisions on sector budgets.

A second explanation points to the role of production contracts in the functional allocation model. These contracts have weakened somewhat the control of the Minister of Health on total expenses for hospital care. It may well be that, obviously for budgetary reasons, the expenditure target for production contracts has been set too low. Delays in the implementation of bed-reduction programs may also cause over-spendings, because the number of beds is reduced less than was assumed in setting sector budgets.

A third explanation focuses upon the presence of some expenditure-raising mechanisms in the functional allocation model which have been overlooked. The merger effect is such a mechanism. If two hospitals merge, the overall budget of the merger-hospital exceeds the sum of the budgets these hospitals would have received if they were still separate entities (the rates for the production items are higher for big hospitals than for small hospitals). Another mechanism is the arrangement that interest costs for loans to cover deficits in operating costs due to delays in payment are subject to full reimbursement. This arrangement does not stimulate hospitals to urge for a quick payment of their full budget.

The gap between sector budgets and hospital expenditures has led to tensions between the Minister and the hospitals, since the Minister has demanded measures to offset the overspendings for hospital care. Hospitals now face cuts in their sector budget. Through increasing their productivity, hospitals are assumed to do more with the same amount of resources.

The policy of budgetary restraint has also affected the financial position of hospitals. Many hospitals have not been able to keep their expenditures within their budget limit. According to a study of the COTG, the proportion of hospitals with a negative result in operating costs was 34% in 1992 (COTG, 1993). This figure highlights how fixed budgets have put many hospitals under great financial pressure. It is thus hardly surprising that they have always associated fixed budgets with expenditure cuts.

THE FUTURE OF FIXED BUDGETS

This chapter discussed some general aspects of fixed budgets in Dutch health care policy with a special emphasis upon the adoption of fixed budgets for hospital care. Over the last decade fixed budgets have rapidly evolved as a cornerstone of modern Dutch cost containment policy, because they essentially put an end to open-ended funding arrangements which were thought totally inconsistent with the goal of effective cost containment. The introduction of the new policy tool in the Dutch health care system incurred great political and administrative complexities, however, since it had been implemented in a health care system that is still mainly financed by a mixture of social and private health insurance. Complexities were further aggravated by the fact that the providers remained to be paid on a fee-for-service basis.

The present prominent role of fixed budgets the Dutch health care policy may perhaps be viewed as another demonstration of an ongoing process of convergence between health care systems. In countries that mainly rely on social and private health insurance in health care finance, the need for cost containment led to the introduction of new policy instruments with which countries of the National Health Service type already have much experience for many years. The widespread use of fixed budgets in the Netherlands has certainly reduced the structural difference in position of the Dutch Minister of Health and, say, his or her British counterpart as regards the opportunities to fix the budgets of the health care providers.

In our view, there is no reason to believe that fixed budgets will be soon given up as a policy tool to achieve cost containment in health care. So far, they have been reasonably effective, although there are some notable exceptions (the fixed budget for specialist care). Perhaps more important may be the fact that the Minister of Finance – who is sometimes depicted as the real Minister of Health – is a fervent supporter of the tool of fixed budgets to constrain the growth of health care expenditure. The Minister of Finance has on several occasions expressed a deeply rooted scepticism of the assumed merits of market competition in respect of health care cost containment, even if it is managed by the government (managed competition). The Minister's policy theory seems to be that market competition will probably only drive up costs instead of reducing them because its basic logic is a strive for expansion and growth. Fixed budgets give much more certainty about health care expenditure than the dangerous adventure of managed competition.

Of course, this all does not mean that there is no room at all for market competition in health care. Providers that operate under a fixed budget will be very interested in market competition if it enables them to purchase goods and services cheaper than without competition. Even the Minister of Finance will favour this type of competition.

We have seen that the introduction of a fixed budget for specialist care has proven to be a tragic failure with even counterproductive effects. It is hardly surprising, therefore, that the Minister of Health is seeking new tools to limit the growth of specialist care and to reduce the probability of deep conflicts with the medical specialists on fee cuts to compensate for budget overruns. One of the policy options, which was also recommended in a report of the Biesheuvel Committee (Biesheuvel was a former Prime Minister of the Netherlands) on the modernization of acute care, is to integrate the hospital budget with the revenues of the medical specialists into a single budget. Hospital management must then internally negotiate with the medical specialists upon the allocation of their revenues. It is unrealistic to believe that this option will receive much support among the specialists, because it essentially means that their revenues become a part of the hospital budget, that they will lose their status of an independent medical entrepreneur in the hospital and that the profitable fee-for-service payment scheme will be abolished.

I have one final comment upon the future scope of fixed budgets. It is not unreasonable to speculate that the scope of fixed budgets will be reduced somewhat in future. Until recently, the distinction between public and private health expenditure was considered largely irrelevant in setting fixed budgets. The current discussion on the introduction of supplementary health insurance as the third compartment in health care finance (the first compartment includes the insurance of necessary long-term care and the second compartment the insurance of necessary acute care) may change the policy landscape, however. The government has stated that it will no longer bear political responsibility for the expenses of health services in the third compartment. If this new view on the government's responsibility for health expenditures politically survives and further materializes in a concrete transfer of more health services from the first and second compartment to the third compartment of supplementary health insurance, then fixed budgets will become more focused upon the public part of health expenditures and to a lesser extent upon the private part. This could be called a fundamental shift in Dutch health care policy making.

REFERENCES

Abel-Smith, B. (1992). Cost containment and new priorities in the European Community. *The Milbank Quarterly*, **3**, 393–416.

Berman, P. (1978). The study of macro- and micro-implementation. *Public Policy*, **3**, 157–84.

COTG (1993). Jaarverslag (Annual Report 1993).

Glaser, W. (1993). How expenditure caps and expenditure targets really work. *The Milbank Quarterly*, **2**, 97–127.

Lieverdink, H. and Maarse, J.A.M. (1995). Negotiating fees for medical specialists in the Netherlands. *Health Policy*, **31**, 81–101.

Maarse, J.A.M. (1995a). Le financement de l'assurance maladie aux Pays Bas.. *Revue d'Economie Financière*, **34**, 325–46.

Maarse, J.A.M. (1995b). Hospital financing in the Netherlands. In Office of Technology Assessment, *Hospital Financing in Seven Countries*. Washington, DC: United States Congress, pp. 195–219.

Maarse, J.A.M. and Kerkhoff, A.H.M. (1995). The politics and administration of Dutch healthcare. In: W.J.M. Kickert and F.A. van Vught (Eds), *Public Policy and Administration Sciences in the Netherlands*. London: Prentice Hall/Harvester Wheatsheaf, pp. 273–94.

Maarse, J.A.M., van der Horst, A. and Molin, E.J.E. (1993). Hospital budgeting in the Netherlands: effects upon hospital services. *European Journal of Public Health*, **3**, 181–7.

Mayntz, R. (Ed.) (1984). *Implementation Politischer Programme II*. Opladen: Westdeutscher Verlag.

Schut, F.T. (1995). *Competition in the Dutch Health Care Sector*. Ridderkerk: Ridderprint.

Schwartz, W.B. (1987). The inevitable failure of current cost-containment strategies: why they can only provide temporary relief. *Journal of the American Madical Association*, **257**, 2, 220–4.

Chapter 7

Fixed Budgets in the Ambulatory Care Sector: the German Experience

Friedrich Wilhelm Schwartz
Reinhard Busse

INTRODUCTION

The introduction of budgetary instruments and their effects in the German ambulatory care sector since the early 1970s presents an interesting case study of the interrelationship between the meso and the macro levels of health care. This German case study can only be understood within the framework of the German health care system, its health care reform process and in particular the cost containment discussion in Germany since the 1970s.

Therefore, we have structured this chapter as follows. First, we will outline the general structure and principles of the German health care system. In this section, we also deal with sectors outside the actual ambulatory care sector on which we will concentrate in the main parts of the chapter. The development from the first voluntary agreements about fixed budgets in the ambulatory care sector in 1976 to their mandatory introduction through the Health Care Reform Act of 1989 is explained in detail. In a separate section, we then analyse both intended and unintended effects of the fixed budgets in the ambulatory care sector. Following this, we outline political interventions to correct the unintended effects and the possible dangers of these corrections. Lastly, we will describe the current situation and possible future developments.

Since reunification took place in the middle of the period discussed in this chapter, all figures presented are limited to the western part of the unified Germany. When analysing the figures, though, one has to consider the fact that Germany as a whole and the health care system in particular were faced with huge problems during the reunification process and the complete re-structuring of the health care system in the east. The fixed budgets discussed in this chapter have to be seen as one solution to cope with these problems.

Fixing Health Budgets: Experience from Europe and North America.
Edited by F. W. Schwartz, H. Glennerster and R. B. Saltman. Technical editor R. Busse.

THE GERMAN HEALTH CARE SYSTEM

The German health care system has a complicated tripartite power structure. It is largely based on the principles of federalism and corporatism which are fundamental principles of German politics inside as well as outside the health care system. Federalism is the basic structural principle of Germany which by constitution is a federal state formed by 16 member states, the so-called *Länder* (eleven of which formed the old west Germany while five were added through unification). Generally, it is intended to hold a substantial part of the power for decisions on the regional level. Corporatism has several important aspects: firstly, it hands over certain rights of the state defined by law to corporative self-governed institutions. Secondly, the corporative institutions have mandatory membership and the right to raise own financial resources under the auspices of, and more or less regulated by, the state. Thirdly, the corporative institutions have the right and obligation to negotiate and sign contracts with other corporative institutions and to finance or deliver services to its members. The contrasting principle to both federalism and corporatism is etatism since the constitution mandates that living conditions shall be of equal standard in all *Länder*. The central state is therefore trying to set uniform rules. The federal *Länder* and the corporative institutions, on the other hand, try to develop their own rules.

In the health care system, the following major actors representing these three principles can be identified. At the central level, the Federal Ministry of Health and, at least from time to time, the parliament are the key players. Federalism is represented mainly by the 16 state governments and – but in fact to a very small extent – by the state parliaments. Corporatism is represented through the legal sickness funds' associations on the purchasers' side and the ambulatory physicians' and dental physicians' legal associations on the provider side. As mentioned before, both sides have mandatory membership. The purchasers' side is made up of 960 autonomous sickness funds (in 1995 as compared to 1152 in 1994 an 1815 in 1970 [BMG, 1995]) which are organized on a regional and/or federal basis. Physicians' associations exist in every state following the principles of federalism. Hospitals, by the way, are not represented by any legal corporative institution. During the discussions of a reform law in 1995, though, the proposal was made to transform the privately organized hospital associations into corporative institutions (Clade, 1995a).

All of these actors have defined missions. Setting the rules and frame for financing health care, equity and comprehensiveness are the missions of the central state level. According to these missions, the German social code book (part V) regulates Germany's comprehensive and compulsory statutory health insurance scheme (Beske, Brecht and Reinkemeier, 1993), especially the following items:

Mandatory and voluntary membership in sickness funds. Membership of a sickness fund is compulsory for employees whose gross income does not exceed a certain level (for 1995: DM 70 200 / year) and voluntary for those above that level. Although the general principles of it have been in place since the introduction of the social health insurance in 1883, the groups for mandatory membership have been increased to dependants, retired persons, farmers, unemployed, students and others over the decades. Currently, 88% of the population are covered by the statutory health insurance: 75% are mandatory members and their dependants, while 13% are voluntary members and their dependants. (Note: 9% of the population are covered by private health insurance, 2% by free governmental health care, for example soldiers, while 0.3% are not insured.)

Frame of the sickness funds' benefit package. This was improved constantly in the 1950s and 1960s (during the *Wirtschaftswunder*) until the early 1970s. The last major steps included full wages during periods of sickness, the final abolition of sickness days without wages in 1969 and the inclusion of preventive services in 1970. Although the Benefits Improvement Act from 1973 which ended limited hospital care coverage, was passed after the beginning of the first oil-crisis, the fact is that this oil crisis marked the end of the golden era of Germany's social welfare expansion. Currently, sickness funds benefits include prevention of disease, screening for disease, diagnostic procedures, treatment of disease, and transportation. Treatment includes ambulatory medical care, dental care, drugs, non-physicians' care, medical devices, inpatient/ hospital care, nursing care at home, and rehabilitation. In addition to these benefits in kind, sickness funds have to give cash benefits to sick insurees after the first six weeks during which employers are responsible for sick pay (amounting to a little over 7% of total expenditure by sickness funds). A third type of benefits are health promotion measures which are offered by sickness funds directly.

Relationships between the sickness funds and providers of health care. Most notably, goals and scope of negotiations between the physicians' and dental physicians' associations and the sickness funds are regulated. Other paragraphs deal with the relationships to hospitals, pharmacies etc.

Organizational structure of sickness funds and their associations. Types of sickness funds include general regional funds (*Allgemeine Ortskrankenkassen*, AOK), company-based funds (*Betriebskrankenkassen*, BKK), guild funds (*Innungskrankenkassen*, IKK), farmers' funds (*Landwirtschaftliche Krankenkassen*, LKK), miners' fund (*Bundesknappschaft*), sailors' fund (*See-Krankenkasse*) and so-called substitute funds (*Ersatzkassen*) which traditionally have been mainly white-collar workers' funds. Until 1995, most blue-collar workers did not have the right to choose their sickness fund. They had to become a member of their regional fund or, if existing, the guild or company-based fund, respectively. White-collar employees, however, had the freedom to choose between the general regional fund and several substitute funds. All funds form legal corporations, have a not-for-profit status and are based on the principle of self-government with elections of board members.

Financing mechanisms. In the chapter of the social code book covering financing, the equal division of contributions to the legal sickness funds between employers and employees and distribution of revenue between the sickness funds according to a "risk-oriented" compensatory mechanism are laid down. All members of a particular sickness fund contribute the same percentage of their income (up to the limit of mandatory membership) which is on average 13.2% (in 1995; Sachverständigenrat, 1995) although actual percentages vary between 9% and 16%.

The *Länder* governments have the mission to secure hospital infrastructure which they try to fulfil by keeping "hospital plans" and paying for investments of hospitals listed in those plans. The investments are paid independently of actual ownership of the hospitals but according to the allocated departments and number of beds. Additionally, the *Länder* are responsible for undergraduate medical education and the supervision of both the physicians' chambers and associations.

The corporative institutions have the full obligation to secure the actual provision of all direct and acute health care services. The most prominent examples of this fact are the physicians' and dental physicians' associations which have both the "Corporate Monopoly and Mission to Secure Ambulatory Care". The main elements of the corporate monopoly and mission are the obligation to meet the health needs of the population, the guarantee to provide state-wide services in all medical specialities and getting a total pre-determined budget from the sickness funds which the physicians' associations distribute among their members (Bogs, 1982; Schwartz, 1995). Because of the absence of corporative institutions in the hospital sector, hospitals contract individually the sickness funds' organizations. Usually, the general regional fund negotiates the contract with the hospital but the conditions, especially the per-diem rates, are valid for all sickness funds. The absence of corporative institutions in the hospital sector on one side and the monopoly for ambulatory care on the other hand is also a major part of the reason for the almost complete separation between the inpatient and the outpatient sectors in Germany.

The tools used by each actor for its specific mission are also different. The main tool of the central government are laws, mainly the social code book V and the reform laws changing and amending it. Included in this legislation is the power for the Minister of Health to issue ordinances. For the state governments, ordinances are much more important. Contracts are the basic tools of the corporative institutions to ensure their mission.

FROM VOLUNTARY TO OBLIGATORY FIXED BUDGETS

The improvements in benefits mentioned were one cause for the steep increase in total health care expenditures between 1970 and 1975: during that time, total health care expenditures doubled, expenditures of the sickness funds increased by 150%, the mean contribution rate rose from 8.2% to 11.3%, and health care expenditures' share of the gross domestic product rose from less than 6% to more than 8% (Sachverständigenrat, 1994). The term "cost explosion" was born and politics started to take a role in controlling health care expenditures. Since then, Germany has experienced an almost permanent debate on cost controlling in health care. This is accompanied by almost permanent legal activity intervening in the financing mechanism of the health care system. Not all of these interventions were cost containment laws in its narrowest sense. Some of them were attempts to regulate the allocation of GP and ambulatory specialist practice locations or to influence the balance between health professional groups with traditionally strong professional corporatism on one side and the legal sickness funds on the other side.

The Phase of Voluntary Agreements in Ambulatory Care 1976–1988

Around 1975 – shortly after the last laws to improve benefits in the social and health care area were passed – the idea of voluntary agreements between providers and sickness funds to control health care costs was born. The federal physicians' association, the *Kassenärztliche Bundesvereinigung* (KBV), as the compulsory organization for every sickness fund affiliated physician and the sickness funds' associations were in 1976 the first to sign such an agreement which limited the growth in the ambulatory care sector to the percentage equal to the rise in incomes from which contribution were made ("contribution income") (Herder-Dorneich, 1982). The actual budgets were fixed for every regional physicians' association (usually identical to the areas of the federal *Länder*) for one year in advance through a contract between that association and the regional sickness funds.

The principle of "revenue oriented expenditure policy" became official state policy through the first health care containment law, the Health Insurance Cost Containment Act (KVKG) of 1977 which followed the 1976 agreement. Through this law, the Concerted Action in Health Care was inaugurated. This marked formally the end of the era of completely free bargaining between supplier groups and sickness funds. In contrast to the voluntary bilateral agreement of 1976, the Concerted Action – including every group of financiers and suppliers in the health care system – had the form of a permanent public consensus conference but overloaded with conflicting interest groups and without sanctions mechanisms. This, in fact, made its task – the definition of prospective growth rates for prices and quantities in all sectors of health care – unobtainable. But its construction at least still followed the illusion of a "negotiated market model".. Another feature of the KVKG was the drafting of a Unified Value Scale (*Einheitlicher Bewertungsmaßstab*) for all sickness funds through which all services were given a point value that could be used both under a pre-determined budget – using the point values as relative weights – and for fee-for-service reimbursement. The Unified Value Scale is determined through a board consisting of seven physicians and seven sickness funds' representatives.

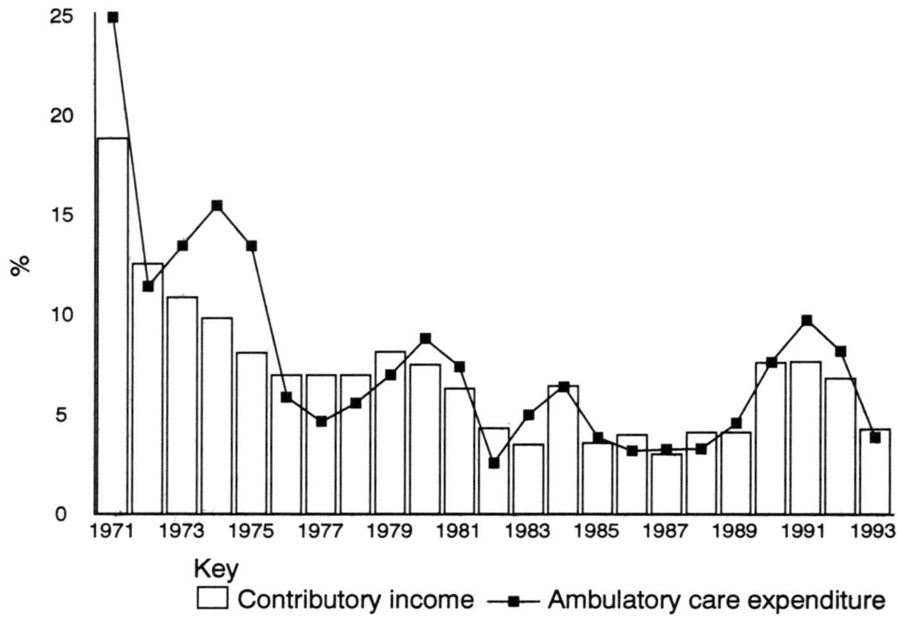

Data source: Sachverständigenrat (1994)

Figure 7.1 Yearly increases in contributory incomes and statutory health insurance expenses for ambulatory care 1970–1993

Through the ongoing annual agreements between physicians and sickness funds, the ambulatory care sector was the only one with remarkably controlled rates of growth (Figure 7.1). Because of greater increases of health care costs in the other health care sectors, the percentage spent on ambulatory care by the sickness funds decreased from 19.4% in 1975 to 17.9% in 1980. Other agreements for dental care and pharmaceuticals did not last; for hospitals and other services no agreement could be reached. The dentists were the first to leave the Concerted Action after only a few months. The white-collar workers' sickness funds were next; they reintroduced "real" fee-for-service charges for their members in late 1978 for some years. The limitation in the ambulatory care sector worked reasonably well throughout the 1980s; the percentage spent on ambulatory care by the sickness funds decreased further to 16.9% in 1988, indicating below average increases compared to other sectors. The physicians' association maintained this limitation, despite protest by many members, with the strategic aim of avoiding a compulsory and legally fixed budget. That this strategy did not work in the final analysis was the result of a lack of control of costs in other health care sectors within the system.

"In the field of ambulatory care there were restrictions in the cost recommendations that can be put down to self-restrictions of the physicians' associations. Thus as far as their performance of control is concerned, the collective negotiations of sickness funds' associations and physicians' associations could be judged as flexible and innovative" (Lauer-Kirschbaum, 1994, pp. 223–4). Döhler (1990, p. 465) states that "collective negotiations of public corporations [are] a secret ideal of the control of resources" in the German health care system. Correspondingly, at the beginning of the discussions on a new reform law, the ministry administration prepared a draft that was based on the idea of self-regulating associations and planned to oblige the statutory health insurance to aim at stable contribution rates (Lauer-Kirschbaum, 1994, p. 237). Of course, it is not known whether in view of recessions the physicians' associations would have succeeded in maintaining this policy of informally balancing out their interests in order to limit growth rates in the long run.

The German Minister of Work, Norbert Blüm, who was responsible for health policy at the time, described the method of self-regulation in a speech before the Concerted Action as follows: "With the help of subsidiarity we prevent our mass society degenerating into a large-scale organization. I regard self-government as a chance to denationalize social policy without privatizing it" (cited in Lauer-Kirschbaum, 1994, p. 232).

The Health Care Reform Act of 1989

Since the cost containment laws did not succeed in limiting the overall growth rate of the sickness funds' expenditures to their contributed income, a major reform law, the Health Care Reform Act (*Gesundheits-Reform-Gesetz*, GRG), was passed in late 1988 and took effect in 1989. The principle aims of this law were a redesign of the health care basket, the reduction in prices of some important items in the basket, the increase of some co-payments to reduce the "moral hazard" elements within the system and a strengthening of the purchasers' side by standardizing and centralizing all negotiating procedures. Blüm's reform law was designed to improve the ability of collective negotiation committees to solve problems especially by adjusting the former imbalance between sickness funds and associations of service-suppliers, which was thought to be structural, by means of further standardization of the sickness funds' decisions. Therefore the GRG introduced a series of obligations for the sickness funds, aiming at a close and uniform co-operation, and it enhanced the status of the federal associations with respect to the regional associations and the single sickness funds. Doing this it ignored the fact that the decisive innovative movements of the last years had been developed by executives from the regional associations of sickness funds and of physicians. Frequently, co-operatively developed problem-solving initiatives had come from innovative regional groups, who tested them practically and gained appreciation and acceptance in the regional committees of physicians and sickness funds by acting as an example or by bringing about formal consensus. The numerous obligations for uniform and joint actions under the GRG, which were introduced with almost no transition period, proved to be a clear-cut obstacle for the implementation of new impulses for and tasks of control that the parliament had included in the GRG.

This reform included, one could say unexpectedly, some additions to the basket, as well. These included the first dental prophylaxis programme for children and youth and, with less scientific background, a screening for cardiovascular and renal disease for adults over the age of 35 years. A major and also costly effort was the interim introduction of home care to be financed by the sickness funds (partly shifted in 1995 to a separate insurance scheme). It is remarkable that these additions to the basket were major elements of a cost containment law. But this was only one side of the coin.

The other side included higher co-payments for dentures, medical devices, transportation and drugs (details in Chapter 8 by Busse and Howorth), but still not for medical services themselves. Most important of all the major changes for the ambulatory care physicians, however, was the legal obligation for the physicians' associations and the sickness funds to fix the budgets for all ambulatory medical services in a way that stability in the contribution rate would be maintained. In fact, the successful strategy of the past became now – against the embittered resistance of the ambulatory care physicians – a legal obligation.

Though the actual size of the budgets still had to be determined through negotiations between the physicians and the sickness funds, it was in fact pre- determined. The only freedom was over decisions about the inclusion of special increases due to innovative services. Both sides used the small remaining freedom, and this resulted in growth rates higher than that of the contribution income. In addition, preventive services were not included in the budgets at all but remained an area for "real" fee-for-service reimbursement which further attributed to the rates of growth.

The fixed budgets did not interfere with the Uniform Value Scale but rather used its points as relative values for rendered services. The conversion factor from points to Deutsche Mark was retrospectively calculated each quarter after the actual volume of services in that quarter was known. In some regional physicians' associations, the over-all budget was (and still is) divided between the different specialities which in effect leads to different point values for different specialities. As a result, physicians do not know the actual fee they will get at the time they provide a particular service.

The hospitals sector was the only one that was ineffectively hit by the law which introduced so-called "flexible prospective budgets", which were negotiated for each hospital individually. The passing of such an ineffective measure for this sector was mainly due to the political resistance of the *Länder*.

EFFECTS OF THE FIXED BUDGETS

The introduction of obligatory fixed budgets in the German ambulatory care sector (and later in other sectors) had several effects. Some of them were explicitly intended and some unintended or at least not explicitly intended by the legislation.

Intended Effects of Budgeting Politics

Intended effects included financial, structural and political effects, i.e. limiting the growth of expenditures, stabilizing the contribution rate, and increasing control over providers. Through the 1989 law, both limitation in the growth of expenditures and thereby the stabilization of the contribution rate were made an obligation for the corporative organizations. Political control over the providers was increased by constituting a centralized framework of benefits for all regions, which removed the power to negotiate regional additional benefits from the regional agencies of the physicians and the sickness funds. The power to negotiate additional benefits was placed with the central associations on both sides instead.

Concerning the first two intended effects, obligatory negotiated budgets between 1989 and 1992 were not as successful as the voluntary agreements. The percentage of sickness funds' expenditure spent on ambulatory rose again above 17% which meant that cost in the ambulatory care sector increased at a little higher speed as total sickness funds' expenditures and at a much higher speed than the contribution income. While in those four years ambulatory care expenditure was up 34%, hospital expenditure was up 33%, total expenditure up 31% and sickness funds' income up 26%. Although the contribution rate had remained stable at around 12.8% during the period from 1989 to 1992 (Sachverständigenrat, 1994), sickness funds were making increasing deficits and had already decided to increase the average contribution rate to 13.4% from 1993. The government decided to act and the Health Care Structure Act was passed (discussed later in this chapter).

Unintended Effects of Budgeting Politics

Several unintended – or at least not explicitly intended – effects can be observed by analysing both utilization and income patterns and political developments.

Intraprofessional allocation conflicts. Fixed budgets had the tendency that allocation conflicts within the physicians' associations and between the different specialities – especially between the general practitioners and the specialists – increased. The main reason for this was the uneven growth of reimbursements (Table 7.1A) and the growing number of physicians (Table 7.1B), which all had to be paid out of the existing budgets. During the time of mandatory but negotiated fixed budgets (1989 to 1992), total income of the physicians grew by 34%, which was well above the inflation rate of 14%. About one-third of this growth was accounted for by the growing number of physicians. About two-thirds of the increase was the result of higher incomes for the individual physicians (Table 7.1C) which was still above the inflation rate. But the growth in incomes varied widely according to speciality with some specialities (i.e. ophthalmology, general practice and internal medicine) barely matching inflation.

Table 7.1 Highest and lowest increases of total reimbursement per speciality / number of physicians per speciality / reimbusement per physician for specialities 1988–1992

Table 7.1A Total reimbursement per speciality		**Table 7.1B** Number of physicians per speciality		**Table 7.1C** Reimbursement per physician for specialities	
Anaesthesiology	+175%	Anaesthesists	+88%	Laboratory specialists	+54%
Laboratory medicine	+ 80%	Neurologists & psychiatrists	+24%	Anaesthesists	+47%
Neurology & psychiatry	+ 54%	Surgeons	+23%	Radiologists	+32%
Surgery	+ 53%	Orthopedics	+18%	Dermatologists	+26%
Radiology	+ 47%	Laboratory specialists	+17%	Surgeons	+25%
Average	*+ 34%*	*Average*	*+11%*	*Average*	*+21%*
Pediatrics	+ 31%	Ophtalmologists	+9%	Pediatricians	+18%
Internal medicine	+ 27%	ENT specialists	+9%	Internists	+17%
Ophtalmology	+ 26%	Internists	+8%	General practitioners	+16%
General practice	+ 23%	General practitioners	+6%	Ophtalmologists	+15%
(Inflation	+ 14%)			(Inflation	+14%)

Data source: Brenner, Heuer and Pfeiffer (1994)

Fee-for-service-payment and the "prisoner's dilemma". The reason for this uneven growth lies in the fact that the remaining two-thirds of the total increase in reimbursement charges is almost entirely the result of a growing number of services, while the value of an individual service was only 0.3% higher in 1992 than in 1988 (Brenner, Heuer and Pfeiffer, 1994). While about one-quarter of the increase in services can be explained through newly introduced benefits (especially through the 1989 reform law), three-quarters may be explained by the "prisoner's dilemma" phenomenon. No matter how the other physicians behave, the individual physician tries to increase his or her share of the total budget by increasing services. Even though this is a rational behaviour for each individual, it contradicts collective rational behaviour which would be to limit the number of services in order to increase the reimbursement per service rendered. On the other hand, the regional physicians' associations as the corporative institutions have been trying to act as the agent for internal collective rational behaviour, limiting extraordinary increases of individual physicians by monitoring their number of services and conducting random audits comparing service frequencies of individual physicians with the average for their speciality.

"Explosion" of technical services. Not all services, though, were suitable for large increases in number. Technical services, especially, could be increased in number much better than time-consuming personal services such as consultations, physical examinations or visits. Technical procedures have the additional advantage that the number of reimbursable points takes investment cost into account. This leads to increasing net gains with increasing utilization since increasing costs are relatively decreasing. Therefore, the number of technical services grew at a much faster pace than average; this is especially true for technical diagnostic procedures, particularly

Table 7.2 Increase in total reimbursement for selected procedures 1988–1992

Procedure		Procedure	
Magnetic resonance tomography	+434%	Consultation	+26%
Histology and cytology	+97%	Physical examination	+24%
Angiography	+90%	Screening for cancer	+18%
Sonography	+85%	ECG	+15%
Computed tomography	+66%	Conventional radiography	+11%
Special laboratory	+66%	Visits	+9%
Average	*+66%*	Basic laboratory	+7%

Data source: Brenner, Hever and Pfeiffer (1994)

imaging (Table 7.2). On the other hand, consultations, physical examinations and visits increased at a slower speed (for some items actually at a rate lower than inflation).

Compensative effects in other sectors? The question of whether fixed budgets in the ambulatory care sector have caused a shift to the hospital sector is difficult to answer since the number of inpatient cases and expenditure was growing even before the voluntary agreements or the fixed budgets were introduced. The speed of that increase has remained constant over time. In terms of sickness funds' expenditure, hospital care accounted for 25% in 1970, 30% in 1975 and 1980, 32% in 1985, 33% in 1990 (Sachverständigenrat, 1994) and more than 35% in 1995.

Inflexibility of the system through the cessation of innovations. In the past, new benefits were usually tried within one region before they were introduced on a country-wide scale. The initiative for innovations came either from the physicians' side or the sickness funds' side. Through the 1989 law with its mandatory benefits package and even more through the 1993 law with the legal fixing of the budgets, the co-operative contract mechanisms became an almost useless and ineffective instrument. Structural innovations – except for some well-defined areas – are no longer allowed and technical innovations are not introduced into the system since the physicians' associations do not want new services within the fixed budgets. They are not willing to accept the same amount of money for a wider range of services but demand an increase of the budgets for theses new services. Due to this conflict, about 40 procedures are currently on a "waiting list".

Eroding the legitimation of the corporative institutions. The GRG diminished the power of the regional associations of physicians and sickness funds and of the single sickness funds by strict procedural regulations and, on the level of the sickness funds, by largely restricting their statutes, which go into matters of the individual fund and into matters of the insured persons. This deprivation of power, however, eroded the basis of authority of these organizations on the spot, especially in terms of pushing through new developments among their own clients or social security partners.

Therefore, the internal authorization and innovative power of self-government did not increase, it deteriorated. This had two main effects. First, it prepared for the partial abandoning of the idea of self-government in the structural health care law. Secondly, the diminishing importance of the co-operative contracting mechanisms have had an effect on the corporative institutions themselves since their basis of legitimation is also diminishing. Discussion is growing – both within the medical profession but also on the sickness funds' side – over whether the physicians' associations should be replaced through free bargaining mechanisms or at least be separated into different institutions for primary care physicians and specialists.

Access for physicians and patients to the system. As the number of physicians was identified as a substantial cost expanding factor in the 1980s, sickness funds and established physicians' associations asked for a stricter limitation of the physician supply through a reduction in the number of medical students and – more relevant in our context – a stricter limitation of physician access to the statutory health insurance system. This contributed to the fact that a growing number of patients consult a specialist directly and not via a general practitioner. Because the average consultation with a specialist is much more costly than with a GP, several models have been politically discussed to establish a more effective or obligatory "gatekeeper function" for the GPs to regulate every access to the system by patients.

POLITICAL INTERVENTIONS TO CORRECT EFFECTS

Since the 1989 reform law failed to control the growth in the sickness funds' expenditure and in the contribution rate mainly because of rising costs in the hospital sector, a new and stricter law, the Health Care Structure Act (*Gesundheits-Struktur-Gesetz*, GSG), was passed late in December, 1992, and took effect almost immediately, at the beginning of 1993. Fixed budgets were – for the first time – extended to the hospital sector. But in contrast to the ambulatory care sector and contributing to the lack of a corporative structure in the hospital sector, an individual budget was fixed for each hospital with a growth rate not higher than that of the contribution income.

The tightening of fixed budgets in the ambulatory care sector was another part of the answer. The regional fixed budgets were now precisely defined by law, removing from the corporative organizations of the physicians and the sickness funds the freedom for negotiations about the sum. The growth of the fixed budgets was now legally determined by the growth rate in the contribution income of the sickness funds' members. Secondly, all services – including innovations and prevention – were included in the budgets (with higher growth rates for certain services, i.e. preventive services and ambulatory surgery). Additionally, the number of sickness fund affiliated physicians was limited to 110% of the average number of physicians per speciality per region. The results of these regulations were twofold: the growth of expenditures was well controlled, but the intra-professional allocation conflict became much more fierce since the number of physicians grew faster than ever before, since limitation of the number physicians was not effective immediately but left a "window" of nine months. In fact, the instrument of limiting the number of physicians led to an increase as big as in the four years previously. This has increased the "prisoner's dilemma" – the number of services is now growing twice as fast as the budget with the result of shrinking reimbursements per service.

To off-set the relative losses in income for general practitioners and to stimulate this sector, fee-for-service payments were supplemented by a nominal additional payment per case for primary care physicians since 1994. These payments which are included in the budgets were made possible by reducing the relative values for laboratory tests and introducing spending – or, rather, point – caps for laboratory tests per patient per physician. A further major revision of the unified value scale was mandatory for the time after 1995; through this revision, commonly used single items such as consulting, ordinary physical examination, and writing a prescription were replaced by a joint "contact fee" with different point values depending on speciality. Additionally, new measures were introduced to lower the profitability of technical services. Above a certain number of procedures per year, fewer points can be reimbursed per service to make up for the decreasing investment costs with an increasing number of procedures.

Another major feature of the GSG was the introduction of fixed budgets in almost all health care sectors affiliated with the ambulatory care sector: To control the prescription volume of physicians, a spending cap was introduced. If spending for pharmaceuticals go beyond that ceiling, regional physicians' associations will be made responsible for the extra costs (see Chapter 8 by Busse and Howorth). Similarly, spending caps were introduced for non-physicians' care and medical devices.

On the other hand, the government's policies also tried to pacify the disconcerted physicians immediately before the 1994 elections by widening their budget by 600 million marks to increase the supplementary fee for general practitioners, but this initiative was stopped by the Social Democratic Party being the majority in the second chamber of parliament, the chamber of the *Länder* governments (Maus, 1994).

Two additional legislative steps in reconstructing the system have taken place. One is part of the above 1993 Reform Act and concerns the traditional strict separation between ambulatory and hospital care. Hospitals may now for some days before and after hospital stays treat patients on an ambulatory basis. In addition, hospitals are now allowed to offer ambulatory surgery. Remuneration for these surgeries is the same as for ambulatory care physicians in private practice. In the first half of 1993, only 4% of all eligible hospitals actually provided ambulatory surgery (Sachverständigenrat, 1995).

An important step towards more active, extended and qualified home care began in

April 1995 with new financial home care benefits (laid down in the social code book part XI). The extension to include nursing home care is planned for 1996. This new long-term care insurance is a new part of the German social network, organized and financed separately but under the same roof as the statutory sickness funds (or by the private insurance companies). However, the income of this new insurance is also financed from employees' deductible payroll payments. It is clear that this new branch of social services in Germany increases the demand for comprehensive and competent ambulatory medical services, and so creates potential for conflict with fixed budgets within ambulatory care.

Beginning in 1996, this change will be completed by offering freedom of choice for nearly everybody insured in the statutory health insurance scheme to choose and change his or her fund (see Chapter 12 by Richard and Schönbach). It is hoped that this will strengthen the competition between funds. On the other hand, through the cost containment Acts in the past the sickness funds as purchaser organizations had been strictly regulated in the sense that every fund offers the same type of ambulatory and hospital services on the same conditions. Competition can, therefore, only rely on better administrative services for the customers or lower administrative costs, both of which seem rather unattractive fields for sickness fund competition. Logically, sickness funds are demanding the possibility of competition in the areas of negotiating benefits and/or organizational structure.

CURRENT SITUATION AND FUTURE DEVELOPMENTS

By law, the era of the legal fixing of budgets in the German ambulatory care sector (and in the hospital sector, as well) was to end by the close of 1995. Now the budgets will again be negotiated between the physicians' associations and the sickness funds. Therefore, to ensure income-oriented expenditures of the sickness funds and a stable contribution rate in the future, a new health system reform has been politically announced for 1996. The other alternative would have been an extension of the fixed budgets but this was ruled out by the governing parties, the Christian Democrats and the Free Democrats, after the 1994 elections.

The Council of Economic and Medical Advisers to the Concerted Action in Health Care (Sachverständigenrat) published an interim report in 1994 and its final recommendations in the summer of 1995. It has been said that the interim report followed the idea of a future "two class medicine". Every insurance model outlined in that report was based on the idea that there had to be a core menu of services covered by the statutory health insurance defined either by the state or by the funds themselves. Around that core, the sickness funds should develop a differentiated menu of tariffs with more choices for consumers. In any case, the consumer would have to pay either for more co-insurance or for more accidental co-payments. Both payment forms are politically described as "more self-responsibility".

In the beginning of 1995, the Minister of Health, Seehofer, announced, though, that he will not accept such a system but rather favours more competition in the corporative negotiation process (Clade, 1995b). The speaker of the parliamentary health commission from the ruling Christian Democratic Union added that harsh budgeting instruments are not planned. In the summer of 1995, the Federal Minister of Health made the proposal to introduce corporative institutions in the hospital sector which then would have the task of controlling hospital expenses through budget mechanisms similar to the ambulatory care sector. In the time of fixed budgets in almost all health care sectors, hospital expenses remained a field with growth rates higher than originally planned and budgeted. In fact, hospitals' fixed budgets grew due to the many loopholes in the budgeting rules. While some hospitals, especially public ones, view the proposal favourably, there is also opposition. This is mainly from the church-owned charity institutions, which do not want to be forced into a legal and ultimately state-controlled corporative institution.

Another proposal, discussed seriously, is to legally fix the contribution of the employers to half of the current average contribution rate and leave a floating contribution share for the employees. In its 1995 expert report, the Council of Medical and Economic Advisors also favours such a freeze (Sachverständigenrat, 1995). In fact that proposal would mean making the insured liable for all difference in health care costs not adjusted by the risk compensation scheme and, probably more important, for all future cost increases. After resistance by both the Social Democratic Party and the unions, Seehofer made the proposal to tighten legal regulations on increases of the employees' part as well (Clade, 1995a) which, in effect, would create a global budget for sickness funds' total expenditures.

The sickness funds, on the other hand, are seeking new areas for competition for new members. They are demanding more competition especially in the area of ambulatory care negotiations with different purchaser models in a more competitive field of providers (see Chapter 12 by Schönbach and Richard). These would then no longer be guided and bound both by legal professional associations and by unified and centralized comprehensive (almost population-wide) contract mechanisms. Generous "open clauses" for alternative purchaser and insurance models of the "managed care" type like HMOs could also mark the end of the traditional system of the uniform population-wide ambulatory and hospital services in Germany. A model with "open clauses" for certain groups in the population would bring new innovative managerial opportunities but also many risks to standards of equity, quality and, at least in the long term, also to cost containment since the uniform, almost population-wide health care provision would no longer exist. But exactly such a social health insurance model including its negotiated allocation mechanisms is believed by many commentators to be the most promising structure for cost-control (Goldberg, Marmor and White, 1995).

In conclusion, the corporative model in the German ambulatory care sector was successful and effective even before the government started to intervene in the negotiation process between physicians and sickness funds in 1989. By reducing governmental budget interventions and increasing the intrinsic freedom for the negotiating partners, the German model could again be a model for the year 2000.

ACKNOWLEDGEMENT

We wish to thank Chris Howorth for his help in preparing this chapter.

REFERENCES

Beske, F., Brecht, J.G. and Reinkemeier, A.-M. (1993). *Das Gesundheitswesen in Deutschland*. Köln: Deutscher Ärzteverlag.

Bogs, H. (1982). Selbstverwaltung im gesundheitsökonomischen Planungssystem der gesetzlichen Krankenversicherung. In H. Bogs, P. Herder-Dorneich, E.K. Scheuch and G.W. Wittkämper, *Gesundheitspolitik zwischen Staat und Selbstverwaltung*. Köln: Deutscher Ärzteverlag, pp. 415–80.

Brenner, G., Heuer, J. and Pfeiffer, A. (1994). *Innovation und Strukturwandel in der vertragsärztlichen Versorgung*. Köln: Deutscher Ärzte-Verlag.

BMG (Bundesministerium für Gesundheit) (1995). *Daten des Gesundheitswesens*. Baden-Baden: Nomos.

Clade, H. (1995a). Seehofers absolute Vorgabe: Beitragsstabilität. *Deutsches Ärzteblatt*, **92**, B1557–8.

Clade, H. (1995b). Ein Warenhauskatalog: Forderung/Wünsche. *Deutsches Ärzteblatt*, **92**, B1658–60.

Döhler, M. (1990). *Gesundheitspolitik nach der Wende*. Berlin: edition sigma.

Goldberg, M.A., Marmor, T.R. and White, J. (1995). The relation between universal health insurance and cost control. *New England Journal of Medicine*, **332**, 742–4.

Herder-Dorneich, P. (1982). Sozialökonomische Entwicklungen im Gesundheitswesen, ihre Auswirkungen auf den ambulanten Sektor und ihre ordnungspolitische Steuerung. In H. Bogs, P. Herder-Dorneich, E.K. Scheuch and G.W. Wittkämper. *Gesundheitspolitik zwischen Staat und Selbstverwaltung*. Köln: Deutscher Ärzteverlag, pp. 133–236.

Lauer-Kirschbaum, T. (1994). Kollektivverhandlungen und Selbstverwaltungskonsens: Interessenegoismus und Gemeinwohlorientierung in der Entwicklung und Reform der gesetzlichen Krankenversicherung. In B. Blanke (Ed.), *Krankheit und Gemeinwohl*. Obladen: Leske und Budrich, pp. 207–44.

Maus, J. (1994). *SPD-Mehrheit im Bundesrat blockiert alle Gesetze*. Deutsches Ärzteblatt 91: B1321–2.

Sachverständigenrat für die Konzertierte Aktion im Gesundheitswesen (1994). Gesundheitsversorgung und Krankenversicherung 2000. Eigenverantwortung, Subsidiarität und Solidarität bei sich ändernden Rahmenbedingungen. Baden-Baden: Nomos (Advisory Council for the Concerted Action in Health Care. Health Care and Health Insurance 2000. Expert Opinion Report 1994. Abbreviated Version in English).

Sachverständigenrat für die Konzertierte Aktion im Gesundheitswesen (1995). Gesundheitsversorgung und Krankenversicherung 2000. Mehr Ergebnisorientierung, mehr Qualität und mehr Wirtschaftlichkeit. Baden-Baden: Nomos.

Schwartz, F.W. (1995). Der Sicherstellungsauftrag für die Kassenärzte und seine Alternativen. *Sozialer Fortschritt*, **44**, 84–9.

Schwartz, F.W. and Busse, R. (1995). Die geplante dritte Reformstufe der gesetzlichen Krankenversicherung – eine Bewertung aus der Public Health-Perspektive. *Medizinische Welt*, **46**, 8–14.

Chapter 8

Fixed Budgets in the Pharmaceutical Sector in Germany: Effects on Costs and Quality

Reinhard Busse
Chris Howorth

INTRODUCTION

The total employment costs of workers in the Federal Republic of Germany, in international comparison, have been a cause of concern for some years. Because the cost of health care coverage for the majority of German employees is a substantial part of the social costs which must be borne by the employer, they have come under close scrutiny and have been the subject of a series of "cost controlling" measures (see Chapter 7 by Schwartz and Busse). In Germany as in most countries, the cost of pharmaceutical products account for a considerable proportion of the total expenditure of the health care sector. Until 1992, pharmaceutical products prescribed in the ambulatory sector accounted for over 16% of the total sickness fund sector expenditure and therefore nearly as much as the expenditure on ambulatory physicians' services (Sachverständigenrat, 1994, p. 347). Sickness funds' expenditure on pharmaceutical products during the 1970s and 1980s rose faster than the contributory incomes of their members (Figure 8.1), the gross domestic product and the general rate of household inflation (Klauber, 1994). At a time of debate over the rising costs in the German health care sector in general, it was not surprising that pharmaceutical expenditure should come under observation. The first cost-containment measures were introduced through the 1989 Health Care Reform Act (*Gesundheits-Reform-Gesetz*, GRG) but more serious changes were introduced through the 1993 Health Care Structure Act (*Gesundheits-Struktur-Gesetz*, GSG), most notably the spending cap for pharmaceuticals, which we will discuss in detail.

In this chapter, we will also examine whether there is any evidence not only for cost containment but also for the hypothesis that the economic rationale for this cap was the lack of incentives to control service efficiency in a system with, as far as the provider of the service and the user of the service are concerned, effectively unlimited

Fixing Health Budgets: Experience from Europe and North America.
Edited by F. W. Schwartz, H. Glennerster and R. B. Saltman. Technical editor R. Busse.

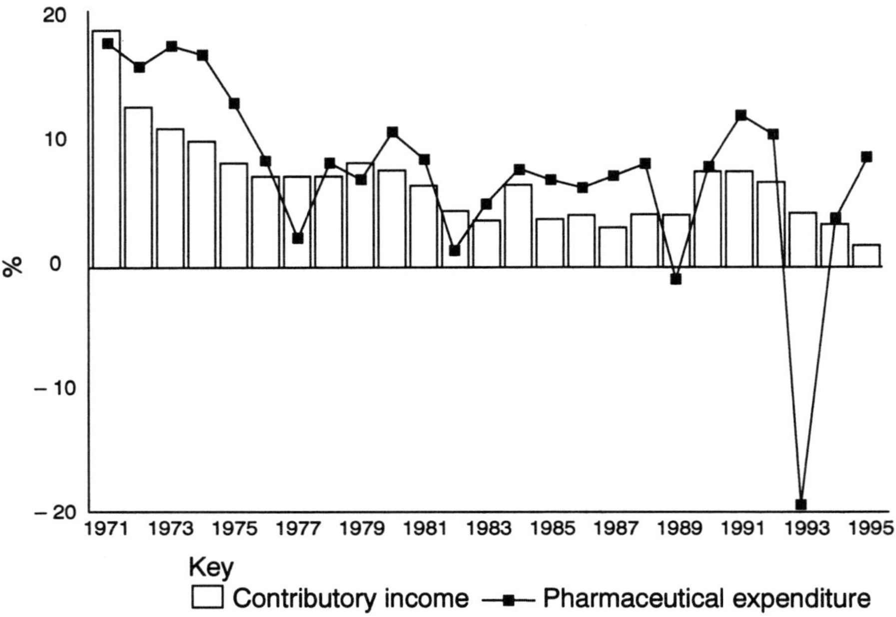

Data sources: Sachverständigenrat (1994); estimates

Figure 8.1 Yearly increases in contributory incomes and statutory health insurance expenses for pharmaceuticals 1970–1995

funding – as in the pre-GSG payment structure. An efficient service should provide its benefits as cheaply as is possible without damage to the intended level of quality. In the case of pharmaceutical prescribing, which previously was a service without direct expenditure constraints, the opportunity cost of efficient service production – i.e. up-to-date knowledge, good communication with the consumer so that he or she is aware of the reason for a particular prescribing decision and the doctor is fully aware of the patient's complaints and wishes or expectations over the treatment process – may well be higher than the self-perceived benefit the practitioner gains from this efficient service production (expressed in levels of patient satisfaction and improvements in clinical outcome observable by the physician). A restriction of the budget as in the GSG should, through the mechanism of (financial) resource scarcity, provide an incentive for medical practitioners to prescribe efficiently. When resources are scarce the consumer of the resources is assumed to derive an increased level of marginal utility from information relating to the relative return on each resource unit used. Therefore, the physician with a restricted pharmaceutical budget and deriving his or her remuneration on the basis of a "fee for service", is assumed to have a strong motive to prescribe in the most cost efficient manner possible, so as to maximize patient satisfaction – and therefore potential future fees – without suffering a present time financial penalty in the form of drug cost repayments. The physician will therefore not prescribe unnecessarily expensive therapies, or therapies which are ineffective. Similarly the physician is unlikely to provide a poor level of service purely in order to reduce immediate costs as this will cause a reduction in future earnings due to patient dissatisfaction.

This chapter is structured as follows. After briefly outlining some important features of the German health care system in respect to pharmaceuticals prescribed in the ambulatory sector, we will concentrate on the first year of the spending cap, i.e. its immediate effects on costs and quality. To this purpose, we will review the available literature, mainly the statutory health insurance's data on the pharmaceutical market (Schwabe and Paffrath, 1994) and four scientific studies (von der Schulenburg, 1994; Güther, 1995; Hoopmann, Schwartz and Weber, 1995; Vitt et al., 1995). Since the spending cap for 1993 was effective only in the western part of Germany, we will limit our results to that area. In the final section we will summarize further develop- ments through 1995 in which we include the eastern part as well.

GERMANY'S PHARMACEUTICAL SECTOR PRIOR TO 1993

The structure of the German health care system is one of corporate, self-governing bodies which operate within a legal framework set out by the federal government and the 16 federal states (*Länder*) which make up the federation. This fusion of corporatism and federalism through the etatist mandates of equity and solidarity throughout the whole German federation is explained in greater detail in Chapter 7 by Schwartz and Busse on the ambulatory care sector in Germany. This structure has led the reforms of the 1970s and 1980s to attempt to control costs largely through the self-governing corporate bodies rather than through (service) allocation ordinances from central government. The tripartite structure has been used in this way to impose centrally developed cost controlling measures in the health sector while leaving the physical resource allocation decisions resulting from these measures largely in the hands of the traditional corporate bodies of the sickness funds and the physicians' associations.

Regulations concerning the pharmaceutical market follow a dichotomy. On the one hand, the distribution of drugs through wholesalers and pharmacies and their respective surcharges on ex-factory prices are regulated in great detail (Schwartz, 1987; Schöffski, 1995). On the other hand, the pharmaceutical industry is remarkably unregulated concerning both pricing and the necessity to prove any effectiveness. The growing realization that a significant proportion of drugs had a level of effectiveness which was unproven and questionable led to the inclusion of the mandate for drug licensing in the 1976 Pharmaceutical Act (*Arzneimittelgesetz*). This regulation is effective only for newly developed pharmaceuticals while the 1994 Pharmaceutical Act Amendment Law extended the deadline for licensing pharmaceuticals already on the market to the year 2005. Therefore, out of the estimated 57 000 pharmaceuticals available in Germany, only 12 000 are approved as effective and unobjectionable in respect to side-effects by the responsible authority (Glaeske, 1994). The structure of the obscure pharmaceutical market was defended by both the pharmaceutical industry and the physicians associations as beneficial for the "therapeutic freedom" of the physicians. Due to this structure, it is not surprising that drugs without any or clear evidence of therapeutical effectiveness are among the most widely sold pharmaceuticals. In fact, Garattini and Garattini (1993) found that six out of the 50 most sold products in 1992 had to be classified in this group which were less than in Italy and France with ten each while for the UK the corresponding figure was zero.

The first step in attempts to reduce the impact of pharmaceutical product expenditure on health care costs was the development of guidelines for reference prices in pharmaceuticals and the introduction of a "negative list" through the 1989 Health Care Reform Act (GRG). The GRG updated all regulations concerning the statutory health insurance scheme with the introduction of part V of the social code book (*Sozialgesetzbuch V*, SGB V). Section 35 of SGB V regulates the establishment of so-called reference prices through negotiations between the corporate associations of sickness funds and physicians. The idea behind the reference prices was to establish an upper limit for the costs reimbursable through the sickness funds. Reference prices were defined in three phases, first for drugs containing the same substance, then for drugs with similar substances and finally for drugs with comparable efficacy. Due to lowered prices for drugs formerly above the reference price, these regulations led to decreasing prices for reference priced drugs but the pharmaceutical industry compensated these through above-average increases for non-reference priced drugs so that the overall effect on costs was very moderate (Klauber, 1994).

For the patients, reference prices had two effects. Generally, pharmaceuticals priced at or below the reference price for that substance were co-payment free (until 1992). On the other hand, if a sickness fund insured patient wishes to use a more expensive alternative, he or she must pay the difference out of their own pocket. For all prescribed drugs without a reference price, the patient had to pay a co-payment of DM 3 per package – instead of DM 2 previously (SGB V, §31). These new regulations led to an increase of co-payments by about one-third, but subsequently – due to the increasing number of reference-priced drugs – it fell to the 1988 level by 1992. While in 1990 reference-priced drugs accounted for only 10% of the drug market, this share increased to about 30% in 1992 and to 40% in 1993 (Klauber, 1994).

These new cost-sharing regulations for pharmaceuticals were part of a complete re-structuring of co-payments through the GRG resulting in generally higher cost-sharing. These included initially 40% to 50% for dentures (SGB V, §30), 10% for non-physicians' care such as physiotherapy (SGB V, §32), DM 20 per ambulance transport (SGB V, §60) and DM 5 per hospital day for up to 14 days (SGB V, §39). Except for co-payments for dentures and inpatient treatment, yearly cost-sharing of mandatory sickness funds' members is limited to up to 2% of their gross income depending on the number of dependants (SGB V, §62). For voluntary members with incomes above the upper limit for mandatory membership, the respective limit is 4% of the gross income. In case of very low income, patients are exempted from virtually all co-payments (SGB V, §61).

Through §34 of the SGB V, a "negative list" containing drugs not paid for by the sickness funds became a legal obligation. The ordinance by the Federal Minister of Health was finally introduced in 1991 and contains medicines used in cases of common flu or constipation. The original list contained 6700 products; 4000 of these were withdrawn from the market. The other 2700 products accounted for about 1.5% of sickness funds' pharmaceutical expenditure.

Another new item introduced by the GRG relevant for the context of this chapter was the extension of the liability of physicians who had general practice expenditures clearly in excess of the average to make repayments (*Regreßpflicht*). Prior to 1989, only physicians whose practice behaviour was widely outside the norm of their specialty had to justify themselves. Practice behaviour included the number of certain diagnostic or therapeutic procedures, referrals to specialists or hospitals, prescribing patterns or issuing of sickness certificates. To prevent a regression to the mean, an additional random sample of 2% of all physicians per quarter was now included in the assessment process (SGB V, §106). Through both measures, around 2% of the total population of physicians are penalized annually (Henke, Murray and Ade, 1994).

These first steps introduced by the GRG did not, however, set any actual, indicative or target budget for pharmaceuticals. This came with the introduction of the 1993 Health Care Structure Act (GSG) which, on the one hand, interfered with the traditional principles of corporatism and federalism more drastically than ever before. On the other hand, it used exactly these structures, especially the physicians associations, to make its regulations work.

REGULATIONS CONCERNING PHARMACEUTICALS IN THE 1993 HEALTH CARE STRUCTURE ACT

The main aim of the GSG concerning the pharmaceutical sector was to contain the sickness funds' pharmaceutical expenditure. The level of containment was, in fact, quite arbitrarily, defined to be 1991's expenditure adjusted for some details mentioned below. By the time of passing the Act late in 1992, it was clear that such a limit could only be kept through a considerable decrease in the number of prescriptions and the value of the prescribed drugs. In fact, in view of the 1992 figure of DM 26.7 billion, the budget of DM 23.9 billion for 1993 meant a reduction of DM 2.8 billion. This was to be achieved through the effects of three measures: price cut/price freeze, new cost sharing regulations, and global ambulatory prescribing expenditure "cap".

Price Cut/ Price Freeze

Article 29 of the GSG introduced a so-called price moratorium for all drugs. Through this regulation, prices for pharmaceuticals with mandatory prescription had to be lowered by at least 5% and for potential over-the-counter pharmaceuticals by at least 2% compared to prices effective on May 1992. Prices for products introduced after that date had to remain constant. Both the price cut and the price freeze were only related to drugs not covered by reference prices and were limited to the years of 1993 and 1994.

New Cost Sharing Regulations

Under the GSG, co-payments for prescribed pharmaceutical (SGB V, §31) were changed in two steps. For 1993, the co-payment rate of DM 3 for non reference-priced drugs and the difference between the sickness fund reimbursable price and the market price of reference-priced drugs were replaced with prescription charges based on the price of the medicines prescribed. Those medicines costing DM 30 or less carried a co-payment of DM 3 provided that the full price of the product was not less than 3 marks, in which case the full price was payable. In the case of medicines priced between DM 30 and DM 50 patients had to pay a co-payment of DM 5, and for medicines priced above DM 50, patients had to pay DM 7. Since January 1994, patients are required to make co-payments on the basis of pack size rather than price, in order to counter the practice of prescribing large packages of medicines which physicians had begun following the introduction of the previous regimes. The current co-payment structure is: DM 3 for small packages, DM 5 for medium-sized packages and DM 7 for large packages. The definition of pack sizes for more than 400 groups is set by an ordinance through the Federal Minister of Health (Bundesminister für Gesundheit, 1993).

Global Ambulatory Prescribing Expenditure "Cap"

Article 29 of the GSG defines a drug budget for the ambulatory sector. This budget is the total amount spent by sickness funds on pharmaceuticals in the ambulatory sector in 1991 (DM 24.4 billion) reduced by:

- 4.25% to cover increased patients' co-payments
- 2.45% because of the mandatory price cut/freeze
- 1.51% due to extension of reference pricing.

The figure produced as a result of these calculations was then increased by:

- 1.74% for price increases in drugs between 1 January 1992 and 30 April 1992
- 0.90% to cover the increase in value added tax from 14% to 15%
- 3.50% to cover the inclusion of additional sickness fund affiliated physicians
- 0.90% resulting from the inclusion of contraceptives in the budget.

The total budget resulting from the above calculation in 1993 was DM 23.9 billion, and in this first year of operation the budget cap applied only to the "old *Länder*" (the former West Germany). This budget was in fact strictly a spending cap rather than a budget in the sense of the global budget for ambulatory physicians described in Chapter 7 by Schwartz and Busse. In the first year the Federal Physicians Association would have been responsible for repaying the first DM 280 million overspent if the budget was exceeded. In this first year, the pharmaceutical industry would have been responsible for any further overspend up to a further DM 280 million – a total therefore between the two groups of DM 560 million.

Four important regulations concerning the spending caps changed at beginning of 1994. First, the pharmaceutical companies are no longer co-responsible for a portion of any overspend and the physicians would be held responsible for the whole of any overspend (without any upper limit) that should arise. Secondly, the five eastern *Länder* were included. Thirdly, the previous federal spending cap was, following historical prescribing patterns, regionalized for all 16 *Länder* forming the unified Germany. Fourthly, the general size of the combined spending caps for all *Länder* may be adjusted through negotiations between sickness funds and the federal physicians' association. Adjustments are allowed for changes in number and age structure of sickness funds' members, drug prices, new legal benefit obligations, increased efficiency and innovative drugs. For 1994, the spending cap for the western Länder was not changed compared with 1993.

One of the features of the tripartite structure of health provision in Germany, mentioned previously in this chapter, is that the regional physicians' associations have the responsibility for deciding how any penalty repayments should be split up amongst their members. Furthermore, because there are no detailed regulations as to how this should be done (the physicians' association of each state may choose its own method) and at the end of the year 1994 it had not been necessary as there had been no budget overshoot, it is still unclear how such redress would be handled. The lifting of the spending cap is permissible only in case the physicians' associations and the sickness funds could agree on physician-specific target budgets taking account of the number, age structure and diseases of patients treated (SGB V, §84). The individual physician would then be liable for prescriptions in excess of 125% of the target (SGB V, §106). Such a regulation would in fact remove part of the corporative power away from the physicians' associations and is therefore rather unlikely.

Although the GSG set a budget cap, expenditure over which would be penalized, there was no mechanism through which savings bringing in expenditure levels under the budget, could be allocated to other areas of ambulatory or, for that matter, inpatient care as is the case in the UK for instance (see Chapter 14 by Glennerster). In addition, there was no mechanism to inform individual physicians of the impact that their individual prescribing patterns had on overall pharmaceutical expenditure, nor were there any immediate quality standards relating to pharmaceutical prescribing.

However, besides the pure cost containment measures concerning the pharmaceutical sector, the GSG did include a regulation to improve the quality of prescriptions and to end the obscurity of the pharmaceutical market. Upon pressure by the Social Democratic Party, §34a was included in the social code book making it legally mandatory for the Federal Minister of Health to issue an ordinance containing all sickness fund-reimbursable pharmaceuticals – a so-called "positive list" – before the end of 1995.

EFFECTS OF THE 1993 REGULATIONS ON COST AND QUALITY

The result of all three cost-containment measures – i.e. price moratorium, new cost sharing regulations and expenditure cap – in their first year of operation was a reduction of 18.8% in sickness funds' costs for pharmaceuticals in the ambulatory sector. In Deutsche Mark terms this figure represents a reduction for the sickness funds of DM 5.1 billion from 1992's expenditure or DM 2.2 billion more than had been required under the terms of the GSG. Of these savings, around DM 1 billion was attributable to price reductions which were 3.9% on average. Another almost DM 800 million were the result of the new cost-sharing regulations; in spite of the smaller market, this meant a total increase of 66.7% in the patients' direct expenditure. Of expenditure for prescribed pharmaceuticals, cost sharing accounted for 7.8% in 1993 instead of 4% in 1992. In other words, only about 60% of the total reduction was attributable to changes in physicians' prescribing behaviour.

Physicians reduced the number of prescriptions by 10.4%. The average value per prescription in 1993 was 4.6% lower than in 1992. Most of this decrease was due to the above-mentioned 3.9% price reduction. The "structural composition" of all pharmaceuticals prescribed attributed the remainder (Selke, 1994). These components include changes between drugs and changes in prescribed strength and package size within one drug. That "structural components" had a negative effect for the first time was largely due to the increase in prescribing generic product instead of the original products. From 1992 to 1993, generic products increased their market share from 30.8% to 35.8% in number and from 24.9% to 28.5% in value (Klauber, 1994).

The above-mentioned increases in patients' cost-sharing do not include their extra costs due to an increase in over-the-counter drug sales. In Germany, almost all pharmaceuticals – except for very few on the "negative list" – may be prescribed by physicians and paid for by the sickness funds – including the ones which may be sold over the counter. In 1992, these drugs were responsible for 37.4% of all prescriptions and 25.3% of the pharmaceutical expenditures in the for sickness funds' market. From 1992 to 1993, prescriptions for (potentially) over-the-counter drugs were reduced by almost 17% in number and DM 1.6 billion or 23.5% in value so that their share dropped to 37.5% and 22.6%, respectively. (On the other hand, prescriptions for drugs with mandatory prescribing were reduced only by 6% in number and 11.4% in value.) A large portion of around 40% of the DM 1.6 billion was substituted by the patients through direct purchases of these drugs resulting in a considerable further cost shifting from the sickness funds towards the patients (Klauber, 1994). As an overall result of these contradictory trends in the over-the-counter market, drugs purchased by patients directly accounted for a larger turnover than prescribed drugs for the first time (Bundesverband der Pharmazeutischen Industrie, 1995, p. 59).

Also, the development in the number of prescribed drugs and daily drug doses differed widely between drug groups on the basis of their respective indications; the highest increase in daily drug doses being 14.6% and the sharpest decrease being 51.7%. Generally, these figures indicate that the decreases for drugs of unproven or disputed effectiveness were on average twice as large as for drugs of proven or undisputed effectiveness. Often-cited examples for the first group are substances prescribed for venous complaints, drugs to stimulate the blood flow, medicines to lower lipid levels, mineral-containing drugs, antihypotensives and vitamins (Table 8.1). Sexual hormones, antidiabetics, beta-blockers, ACE-blockers and corticosteroids, on the other hand, were prescribed more often in spite of the spending cap. These trends are viewed by many politicians and sickness funds' administrators as proofs for a rational behaviour of the physicians and against any decrease in prescribing quality (Selke, 1994). It has to be taken into account, however, that a majority of unproven or disputed drugs are available as over-the-counter drugs which, as noted above, increased their market share drastically. There is evidence that patients did not switch voluntarily but were asked to do so by their physicians who no longer prescribed drugs "requested" by the patients as generously as before (Glaeske, 1993).

The data concerning whole groups of indications have the disadvantage of being aggregate data without any information about individual physicians or patients. Therefore, they allow only limited conclusions about the actual quality of physicians' prescribing patterns. Unfortunately, the introduction of the relatively harsh cost containment measures in 1993 was not accompanied by research stimulated or even financed through federal institutions. Therefore, we have only a limited number of studies which evaluated various aspects of ambulatory physicians' prescribing behaviour in the first year of the GSG's pharmaceutical measures. These studies indicate effects both for the costs and quality of pharmaceutical prescribing practice which are usually overlooked.

A study of 76 physicians in Schleswig-Holstein (Vitt et al., 1995) analysed the

Table 8.1 Changes in defined daily doses (DDD) for 1983–1992 and 1992–1993 (selected groups with at least 100 million DDD; western part of Germany only)

	1992 vs. 1983 %	1993 vs. 1992 %
Sexual hormones	+361	+14.6
Antidiabetics	+24	+5.3
Beta-blockers/Calcium-antagonists/ACE-blockers	+226	+4.8
Corticosteroids	+107	+3.9
Diuretics	+71	–2.0
Antihypertensives	–24	–7.7
Broncholytics/antiasthmatic drugs	+136	–7.9
AVERAGE	+46	–9.3
Gastro-intestinal therapeutica	+44	–14.9
Vitamins	+97	–17.9
Antihypotensives	–29	–22.3
Mineral-containing drugs	+261	–22.8
Substances to lower lipid levels	n.a.	–23.7
Substances to stimulate the blood flow	+29	–26.5
Venous/ antivaricose therapeutica	+25	–33.3

Data source: Klauber, 1994; Selke, 1994

prescribing behaviour of 16 physicians, selected because they had been identified as those with the highest level of prescribing expenditure within the group. The prescriptions of this group were analysed on the basis of the accompanying diagnosis by a clinical pharmacologist with the aim of assessing changes in prescribing practice in terms of cost, quantity and quality following the coming into force of the GSG's drug budget cap. The prescriptions of these 16 physicians in the last quarter of 1991 were compared with those in the first quarter of 1993. These physicians constituting the upper cost quartile reduced the number of prescriptions by a considerably higher margin than the average German physician, i.e. by 27%. The costs of the prescribed drugs fell by 29.8% indicating that there was no trend towards cheaper pharmaceuticals (since the general price reductions have to be taken into account).

To assess the prescriptions in terms of quality, four categorizations were used:

1. Therapies of proven benefit for the given diagnosis, which would also be considered to be the standard treatment for this diagnosis.

2. Therapies of proven benefit for the given diagnosis, but for which other equally effective but cheaper therapies exist; for example: therapies with ingredients that were more potent than indicated, new pharmaceuticals which are indicated only in exceptional cases or use of "original preparations" in spite of an existing suitable generic product.

3. Therapies which are obsolete on grounds of outdated ingredients, risk of dependency, high risk or other unsuitability.

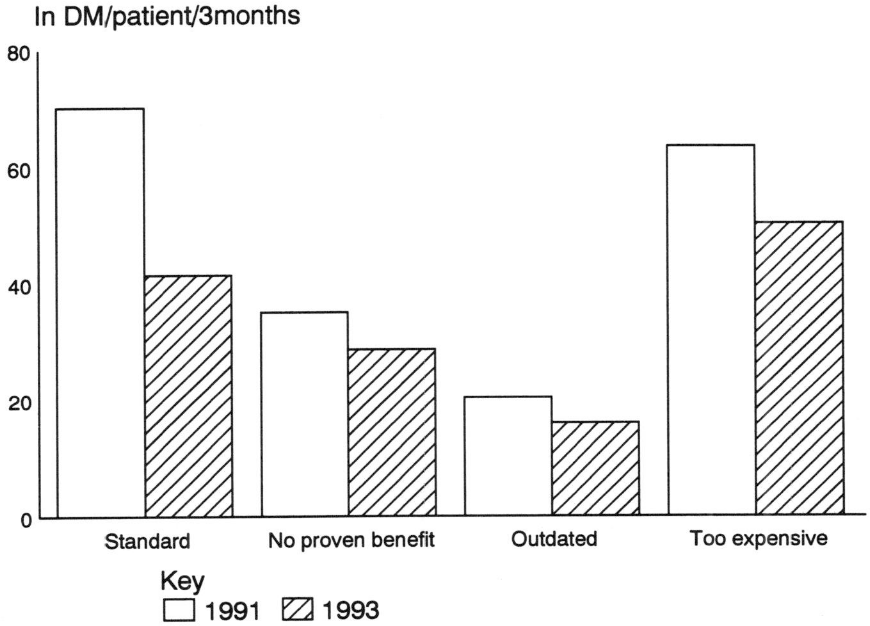

Data source: Vitt et al., 1995

Figure 8.2 Changes in prescription volume for pharmaceuticals grouped according to therapeutic quality

4. Therapies of unproven benefit: These may be useless therapies or they may be therapies which are possibly beneficial but have not undergone clinical trials, usually due to having been introduced before such trials were legally required by the *Arzneimittelgesetz* passed in 1976.

After categorizing each prescription written by the 16 practitioners in the two periods it was discovered that those physicians did display a high level of inappropriate prescribing behaviour. This was evidenced by the high level of over-expensive treatments, outdated treatments, and treatments with unproved therapeutic benefit (see Figure 8.2). However, when the prescribing behaviour of these physicians after the advent of the budget cap is observed it was evident that although the physicians reduced their overall number of prescriptions and the average value of each prescription, the largest decrease was that in appropriate therapies, although only a relatively small percentage of their prescriptions fell into this – "indicated therapy" – category (37% in 1991 and 30% in 1993). Those therapies in the other three categories, and especially those in the second "too expensive therapy" category (34% and 37%, respectively) actually accounted for a *higher* proportion of the total drugs prescribed after the GSG. In other words, although prescribing in all four groups was reduced, the quality/efficiency outcome of the GSG is potentially negative. As stated above, it may be argued that prior to the GSG physicians tended to prescribe too liberally even in the case of "standard" therapies, and that therefore a reduction in prescribing causes, at the least, no measurable decline in treatment quality (and could possibly lead to an improvement in quality). However, while it is not yet possible to draw any outcome conclusions over the public health impacts of the GSG the increase in the proportion of "un-indicated" pharmaceutical therapies is certainly a negative development, and it is highly possible that there has been a reduction of indicated prescribing which could result in a decline in public health in the long term.

There may be more than one reason for poor prescribing behaviour, such as:

1. The physician may have a patient mix with a higher level of morbidity or more serious types of morbidity than the average; for example, an unusually high number of diabetic patients.

2. The reason for poor prescribing practice may be patient demand. Patients who expect to be given a prescription when they visit a physician quite simply "buy" the services of physicians willing to prescribe "to demand" rather than those of a physician with stringent prescribing criteria. In this case poor prescribing practice is "market driven".

3. The physician may find it easier or personally more economically efficient to be a high "prescriber". Issuing prescriptions, and especially repeat prescriptions not requiring an actual consultation, leads to a higher and quicker throughput of patients.

4. The physician is more easily influenced by the marketing of the pharmaceutical industry or has a lower threshold of willingness to try new (and usually more expensive) therapies.

5. The physician simply does not have a high enough awareness of pharmacology and is *unable* to make the correct prescribing choices.

Any one or combination of these factors may cause the prescribing outcomes seen above. The worrying fact remains that the capping measure did not really improve the quality of the prescribing behaviour.

A second area for concern is the possibility that physicians will attempt to maintain patient satisfaction levels not by changing patient expectations of prescription levels, but by referring patients to other speciality groups or into the "inpatient" sector. In view of the exclusion of the "inpatient" sector from specific pharmaceutical budgeting measures, increasing referrals would protect ambulatory physicians from penalties under the terms of the GSG but leave the total drug expenditure of the sickness funds unchanged. Although the *Regreßpflicht* sanction, outlined earlier, may be considered to provide some disincentive for excessive use of this tactic – but as an incentive to refer patients from high prescribing practices to low prescribing practices – it has to be borne in mind that if the majority of physicians in a particular speciality revert to this practice this change will then be seen as the "norm" and as a result the rising rate of referrals will not lead to any doctors being penalized as "outliers". The total cost of treatment resulting from such changes in referral practice might even be higher, due to the higher cost structure of treatment by specialists resulting from their higher level of technical intervention and the much higher cost of "inpatient" treatment which must include "hotel costs" as well as treatment costs owing to the lack of hospital "outpatient" sector in Germany.

A study by von der Schulenburg (1994) at the beginning of the period covered by the pharmaceutical budget cap did indeed measure a significant increase in the level of referrals made by a panel of more than 300 physicians whose referral behaviour was compared for the years 1992 (before the coming into force of the budget cap) and 1993 (after the imposition of the cap). The increase in referrals led to significant increases in hospital admissions; von der Schulenburg has estimated that possibly more than half of the DM 5.1 billion saving resulting from the GSG's drug cap was spent as a result of direct and indirect costs associated with this change in referring practice (not considering that the ambulatory care sector has a fixed budget). This calculation is without consideration of the "external costs" resulting from increased patient anxiety etc. and does not take into account the impact of overall changes in prescribing behaviour on service quality, although he states that "it is likely that the increased prescription of cheaper and older drugs leads to poorer quality". Additionally, as a result of the (admittedly ineffective) budgeting measures for overall hospital expenditure tying increases in hospital expenditure to the rate in rise of contributory income, it can be expected that an increase in referrals from the ambulatory sector may have a "crowding out" effect on hospital service provision and, therefore reduction in overall service quality.

A study of 82 general practitioners and general internists with almost 3 million patient contacts in 1992 and 1993 (Güther, 1995) clearly confirms the findings of von der Schulenburg in the first quarter of 1993. The analysis of Güther's figures reveals that in the first quarter of 1993 there was an increase of 20% over the corresponding period in 1992. The result was that in each quarter the increase in the number of referrals per hundred consultations fell until in the final quarter there was no increase in the number of referrals compared with the same period a year before (per quarter: +20%, +11%, +9% and +0%). The overall year on year increase was 10%. This result would suggest a "panic" reaction in the first months of the legislation's coming into effect, with a settling down process and finally a return to pre GSG referral patterns with a rate of referral rising only slowly, a trend which was already established prior to the GSG.

Perhaps more relevant is the trend which is exposed through the breakdown of prescribing figures into two "raw" cost groups. Güther's study divides the practitioners in the sample into two equal sized groups, one group containing the practices with the highest prescribing costs, the other the practices with the lowest costs. Prescribing by these two groups was then compared year on year over the eight quarters of 1992 and 1993. The change in prescribing practice is much more marked in the expensive practices than in the cheaper practices. The total year on year reduction between the two groups together was 6%, but when the two groups were separated those physicians with the higher prescribing costs reduced their number of prescriptions per patient contact by 9% while those in the bottom half of previous prescribing expenditure decreased their number by only 3%.

Considering the findings of the Vitt et al. study discussed above and assuming that practices with the more expensive prescribing are those with the poorer prescribing quality, it may be concluded that physicians with poor prescribing decisions *are* reducing their prescribing costs more steeply than those who have better prescribing patterns (providing the base assumption that the most expensive prescribers *are* indeed the poorer prescribers is correct), but they are *not* improving their prescribing decisions. It should also be considered that if it *is* assumed that those doctors who already had prescribing expenditures in the lower half of the scale had achieved this through the application of good prescribing practice any reduction made by this sector (albeit a small reduction) may be damaging for the long-term health outcome of their patients. In addition, the effect of the GSG is – as were previous legislative measures – short term in that the tendency is for an initially strong reaction to the cap with a subsequent tendency for a return to pre GSG prescribing practices. The result of this combination of factors is a risk of poorer quality overall without a long-term impact on the trend for pharmaceutical prices.

Finally, a study by our own department about the effects of the GSG on the medical treatment of four common health problem groups treated by ambulatory physicians should be considered (Hoopmann, Schwartz and Weber, 1995): 130 physicians operating in a sentinel network (a 37% sample of 350 randomly selected general practitioners and internists from Lower Saxony) were surveyed for both their opinions and actual prescribing behaviour over the changes in the treatment. Concerning patients who had been treated in the same practice for the same complaint prior to the GSG drug expenditure cap, the physicians were asked at four different stages during 1993 to document the number of and reasons for changes in treatment of their patients. The physicians were then asked for their opinion of the impact these changes made to the clinical quality of the treatment and the impact of the changes on patient satisfaction. The four surveys, each covering a different topic, were taken in May, June, October and November. Therefore the study results are based on the analysis of four cross-sectional surveys recording the effects of the GSG in its first year of operation and indicates the level of adjustment to the law over time. In the first group the physicians reported a change in treatment in 27% of cases. This level of treatment change fell in subsequent groups to 22%, 18% and 16% respectively. Only a minority of changes were attributed to the GSG (11%, 5%, 3% and 2% of the total cases respectively) and it can be seen that the proportion of changes attributed to the GSG fell sharply throughout the period of the study. Some of the fall in proportion of change attributed to the GSG may be associated with a cognitive attempt by the physicians to deny willingness to change treatment on purely financial grounds. It does, however, also correlate with the assumption of an initial "panic" reaction to the GSG and subsequent return to normal practice seen in the analysis of Güther; the non GSG level of change being relatively steady over time around 15%.

The main findings of the study were that physicians, when asked to assess retrospectively the GSG-related changes in the beginning of 1993, expected negative effects on patient satisfaction in more than 40% of cases, on therapeutic effect in 30% and potential side-effects in 25%. Later, they expected such bad outcomes only rarely, probably indicating a decreasing willingness to accept these – corresponding to the decrease in GSG-attributed changes. On the other hand, the doctors' opinion on patients' satisfaction following GSG-based changes in treatment indicated a remainder of expected dissatisfaction, especially in cases viewed by the patient as severe and by the doctor as minor. In contrast, treatment changes not caused by the GSG were assessed as increasing patient satisfaction. It is likely that this finding is to some extent distorted by the physicians' feelings of dissatisfaction and loss of autonomy following the GSG, and a desire to be able to continue making clinical decisions without the need to keep "looking over his or her shoulder" to ensure compliance with a budget restriction. However, ambulatory doctors in Germany are likely to be acutely aware of patient satisfaction as a result of the freedom of patients in the German health care system to move freely from doctor to doctor.

As might be expected from the result of the Güther study, Hoopmann, Schwartz and Weber found that physicians who had previously been penalized financially (*Regreßpflicht*) because they had not conformed to "economic prescribing" changed their treatment strategy almost 3.5 times as often as their colleagues to achieve cost containment. Although the treatment groups in the study were different in each of the serial cross-sectional surveys, the similarity of the level of non GSG changes together with the unidirectional drop off in GSG influenced changes and the possibility of an increase in referrals to the "inpatient" sector identified by von der Schulenburg indicate reliability in the conclusion of long-term inefficacy of the cost containment measures contained in the GSG.

When considering the impact of the GSG changes on the quality of health provision the opinion of the physicians on the clinical quality effects of the changes may lead to a level of complacency over the quality impact of the changes wrought by the GSG in prescribing patterns but until it is possible to ascertain the long-term direction of outcomes following these changes it is not possible to be certain that quality is unaffected by these changes. In the meantime – and especially as patient satisfaction levels are both currently important for German doctors economically and seen as an

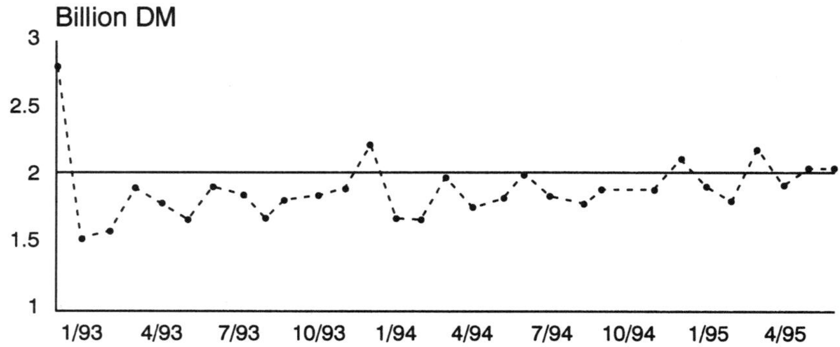

Figure 8.3 Actual monthly sickness funds' pharmaceutical expenditure as compared to the (hypothetical) monthly spending cap December 1992–May 1995

important area for future quality measurement – expressions of patient dissatisfaction with the results of the GSG should be taken seriously. Further it is likely that patient compliance with a treatment regime with which they are unhappy, will be lower than with one with which they are satisfied. It is also likely that in the long term high levels of dissatisfaction in the physicians' professional circle may lead to a loss of motivation, and problems in future negotiations between the physicians and the sickness funds similar to those already experienced in the dental sector. Such problems could potentially lead to a loss of sickness fund practitioners and subsequent endangerment to levels of public health. These issues should be seriously considered when making amendments to the structure of the health sector.

FURTHER DEVELOPMENTS

In 1994, sickness funds' pharmaceutical expenditure increased by 3.9%. In spite of this increase, it remained on average 5.4% below the spending cap in the western *Länder* (Figure 8.3). In the east, however, there was an overspending of 2.4% on average (Krimmel, 1995) which had no immediate consequences since an overspending in one year may be evened out in the following year. Evaluating 1994's figures, it has to be taken into account, however, that due to the new cost-sharing regulations, patients' cost share increased by about one-quarter compared to 1993. After the end of the two-year price moratorium, pharmaceutical prices started to rise again in early 1995, especially for drugs without reference price. In January 1995 alone, they became 1.9% more expensive. This was contributing to the fact that pharmaceutical expenditure in the first half of 1995 was about 10% higher than in 1994 (Figure 8.3) which would result in repayments to the sickness funds if the spending cap had remained at the 1994 level. The federal physicians' association was therefore speaking of a "threat to the economical livelihood of the physicians"

(Krimmel, 1995).

The GSG's aim, a reduction in sickness funds' pharmaceutical expenditure, could have been achieved through the exclusion of all pharmaceutical products with no proven benefit. The Wissenschaftliches Institut der AOK (1995) has calculated that the proposed "positive list" would lead to a saving of at least DM 2 billion – the (unlikely) substitution of all excluded products with recognized substitutes taken into account. This substitution effect was calculated to be DM 5 billion. This possible reduction could be expected to increase the outcome quality of the health provision system purely on the basis of the use only of useful medications. In addition, the financial savings would in all likelihood be greater than the two billions calculated, as it is unlikely that other medicaments would be substituted in all cases. Control over the rate of increasing pharmaceutical costs, could be achieved by continuing education of physicians and more involvement of pharmacologists in prescribing decisions without a budget cap. Such measures could also be reinforced through self-regulation by the physicians' self-regulatory bodies with the aid of sample checks on doctors' prescribing and targeted training (with possible sanctions for persistent poor practice) for those practitioners found to be poor "prescribers". Practising physicians seem to appreciate the help of drug formularys (Kochen, Sondholzer and Himmel, 1994) which in effect a "positive list" would be.

The Federal Minister of Health, though, decided not to pursue the idea of a "positive list" and proposed a change of the social code book – the abolishment of §34a. He justified this with the successful cost containment measures in the pharmaceuticals sector, otherwise rising costs for chronic patients due to over-the-counter purchases and, most importantly, the threat to smaller pharmaceutical companies. While this decision was welcomed by the pharmaceutical industry, it was faced with criticism by both sickness funds and the Social Democratic Party. But no matter whether the law would be passed or not, the "positive list" was dead by the summer of 1995. In the case of the latter, the Federal Minister of Health threatened to include all pharmaceuticals currently on the market in the list.

CONCLUSION

The introduction of sharp cost containment measures in Germany's pharmaceutical sector, i.e. a price cut for pharmaceuticals, higher co-payments by the patients and a spending cap on sickness funds' pharmaceutical expenditure, led to an 18.8% reduction in sickness funds' pharmaceutical expenditure in the first year. However, this effect did not last since physicians started to increase the number of prescriptions again in 1994 and markedly in 1995 in spite of the threat to be held financially responsible for any overspending. The pharmaceutical industry started to increase prices in 1995 after the end of the forced price moratorium. The only lasting change was the rise in patient cost sharing.

In spite of an initial disproportionately high drop in prescriptions for drugs of unproven or disputed effectiveness, no improvement in the quality of the prescribing behaviour could be observed when results of scientific studies about effects during the first year of the cost containment measures were taken into account. Therefore, we agree with Soumerai et al. (1987) who, in an article of about payment restrictions for prescription drugs in New Hampshire, stated: "*Primum non nocere* (first do no harm) is one of the principle tenets of medical practice. It is not unreasonable to expect health policy makers to assess the potential unintended harm to patients that may result from previously untested innovations." The evidence above suggests that the enactment of the GSG has not followed this maxim. During the discussion for a further Health Care Reform Act in 1996, the Minister of Health, Horst Seehofer, announced that in the future drastic changes would be introduced only when they had been subjected to carefully evaluated trials.

REFERENCES

Bundesminister für Gesundheit (1993). Verordnung über die Zuzahlung bei der Abgabe von Arznei- und Verbandmitteln in der vertragsärztlichen Versorgung. Re-printed in: Bundesverband der Pharmazeutischen Industrie (Ed.) (1994), *Rote Liste 1995*. Aulendorf: Edition Cantor, pp. 405–12.

Bundesverband der Pharmazeutischen Industrie (1995). *BPI-Jahresbericht 1994*. Frankfurt: Bundesverband der Pharmazeutischen Industrie.

Garattini, S. and Garattini, L. (1993). Pharmaceutical prescriptions in four European countries. *The Lancet*, **342**, 1191–2.

Glaeske, G. (1993). Die Budgetierung der Arzneimittelversorgung – eine Herausforderung für die Qualitätssicherung. *Die Ersatzkasse*, **73**, 45–51.

Glaeske, G. (1994). Behindern die gesetzlichen Regelungen eine ausreichende Qualitätssicherung im Arzneiversorgungsbereich? (Abstract in English: Do legal regulations obstruct good quality assurance in the field of drug therapy?) *Das Gesundheitswesen*, **56**, 80–5.

Güther, B. (1995). Auswirkungen des GSG auf das Verordnungsverhalten niedergelassener Ärzte. (Abstract in English: Effects of German health legislation on the prescription of GPs.) *Das Gesundheitswesen*, **57**, 185–91.

Henke, K.D., Murray, M. and Ade, C. (1994). Global budgeting in Germany: lessons for the United States. *Health Affairs*, **13**, 4, 7–21.

Hoopmann, M., Schwartz, F.W. and Weber, J. (1995). Effects of the 1993 health reform law upon primary care practitioners individual performance: results from an empirical study in sentinel practices. *Journal of Epidemiology and Community Health*, **49**, Supplement 1, 33–6.

Klauber, J. (1994). Entwicklung des Fertigarzneimittelmarktes 1983 bis 1993. In: U. Schwabe and D. Paffrath (Eds), *Arzneiverordnungs-Report '94*. Stuttgart-Jena: Gustav Fischer Verlag, pp. 465–89.

Kochen, M.M., Sandholzer, H. and Himmel, W. (1994). Attitudes of primary care physicians towards the use of a drug formulary – preliminary results of a study in Germany. *International Journal of Clinical Pharmacology and Therapeutics*, **32**, 400–2.

Krimmel, L. (1995). Existenzbedrohung für die Kassenärzte. *Deutsches Ärzteblatt*, **92**, B736–9.

Sachverständigenrat für die Konzertierte Aktion im Gesundheitswesen (1994). *Gesundheitsversorgung und Krankenversicherung 2000. Eigenverantwortung, Subsidiarität und Solidarität bei sich ändernden Rahmenbedingungen*. Baden-Baden: Nomos.

Schöffski, O. (1995). *Die Regulierung des deutschen Apothekenwesens – Eine ökonomische*

Analyse. Baden-Baden: Nomos.

Schulenburg, J.-M. von der (1994). Überweisungs- und Einweisungsverhalten von Primärärzten nach dem GSG. *Münchener medizinische Wochenschrift*, **136**, 202–5.

Schwabe, U. (1994). Überblick über die Arzneiverordnungen im Jahre 1993. In: U. Schwabe and D. Paffrath (Eds), *Arzneiverordnungs-Report '94*. Stuttgart-Jena: Gustav Fischer Verlag, pp. 1–16.

Schwabe, U. and Paffrath, D. (Eds) (1994). *Arzneiverordnungs-Report '94*. Stuttgart-Jena: Gustav Fischer Verlag.

Schwartz, F.W. (1987). Arzneimittel und Apotheker im Zeichen von Kostendämpfung und Strukturreform im Gesundheitswesen. *Pharmazeutische Zeitung*, **132**, 2801–10.

Selke, G.W. (1994). Auswirkungen des Gesundheitsstrukturgesetztes. In: U. Schwabe and D. Paffrath (Eds), *Arzneiverordnungs-Report '94*. Stuttgart-Jena: Gustav Fischer Verlag, pp. 447–64.

Soumerai, S.B., Avorn, J., Ross-Degnan, D. and Gortmaker, S. (1987). Payment restrictions for prescription drugs under Medicaid: effects on therapy, cost, and equity. *New England Journal of Medicine*, **317**, 550–6.

Vitt, K.D., Buss, G., Erben, C.M. and Schönhöfer, P.S. (1995). Beeinflußt eine Budgetierung Qualität und Wirtschaftlichkeit der ambulanten Arzneimitteltherapie? (Abstract in English: Will a predetermined budget influence the quality and efficiency of ambulatory drug therapy?) *Das Gesundheitswesen*, **57**, 171–6.

Wissenschaftliches Institut der AOK (1995). *Vorschlagliste für Positivliste für Arzneimittel*. Bonn: Wissenschaftliches Institut der AOK.

Chapter 9

Federal and Provincial Roles in Canadian Health Care Budgets

Lawrence J. Nestman

INTRODUCTION

For many Canadians, their national public health insurance system is a significant social, cultural and economic achievement (Taylor, 1987). However, while modern health technology has brought many advances to the Canadian population, increasing costs have taxed the capacity of governments to fund the level of health services to which Canadians have become accustomed. In addition there are concerns about access to care, at a time when knowledgeable observers are contending that a significant share of existing health care expenditures are being spent on unnecessary and inappropriate services. As a result Canada is reappraising the organizations designed to provide health services and is making significant changes in the way services are delivered, administered and financed.

HISTORICAL REVIEW OF SOME OF THE BASIC CONCEPTS, PRINCIPLES AND FUNDING MECHANISMS THAT UNDERLIE THE CANADIAN HEALTH CARE SYSTEM

The Early Years to the End of the 1960s – A Period of Growth

In the early years of this century the delivery of health services was a local affair, usually the responsibility of local governments or the church (Nestman, 1995). When it became clear that local administrations did not have the financial tax base to fund health services, the public looked for new arrangements for funding. The vast distances, climate, poor transportation and communication systems contributed to access and service delivery problems. By the 1940s, the need for health insurance became evident. Throughout the 1940s and 1950s both the provincial and the federal governments were increasingly pressured to provide direct funding for health care.

Fixing Health Budgets: Experience from Europe and North America.
Edited by F. W. Schwartz, H. Glennerster and R. B. Saltman. Technical editor R. Busse.

This pressure took the form of increased evidence of the private sector's limited ability to provide an adequate level of service coverage combined with pressure from World War II veterans (who expressed the view "We and our families had coverage during the war, why not now?") and from immigrants arriving in Canada from countries with existing social health systems such as the UK and Germany.

The leadership role in the development of health services initially took place in 1943, when the Saskatchewan government enacted North America's first compulsory state sponsored universal hospital insurance program. The Federal government followed this lead in 1958 with the financing of a similar program on a 50–50 cost-sharing basis with all the provinces. Provincial governments became the main payers of costs incurred under a retrospective, cost-based payment system with a line by line budgeting method (Nestman, 1989).

In 1962 the Saskatchewan government enacted North America's first compulsory state sponsored universal medical care insurance program. Once again the Federal government followed Saskatchewan's lead and in 1968 agreed to finance a similar program on a 50–50 cost sharing basis with the provinces. Physicians were paid on a fee-for-service basis utilizing a fee schedule agreed upon between the medical profession and the provincial government (Nestman, 1989). Physicians were allowed to bill over and above the agreed schedule. In 1966 the federal government passed the Canada Assistance Plan sharing the costs of social services for low-income persons on a 50–50 basis with provinces. The Canada Assistance Plan helped fund the development of the welfare state in Canada.

Because the payments under the cost-sharing agreements were based on a national average, the wealthy provinces received from the federal government a lower percentage of their actual costs than the lower income provinces received. In this way, the federal government became a social leveller and standard setter (Nestman, 1970). As a commitment to this view, provincial governments increased the number of hospitals with the belief that the best care that could be provided was in the acute care hospital.

In the latter part of the 1960s, feeling that under the cost-sharing agreement it was required to pay a share of growing costs that were generated solely by the provinces the federal government set up a task force to study these costs (Government of Canada, 1969). The task force recommended the use of the global budget method, which is now the most common method of funding used in Canada (Nestman, 1989). This new system is, in contrast to the old retrospective system, prospective and therefore provides an incentive for greater efficiency.

Developments in the 1970S and 1980S – A Period of Cost Containment

Established Program Financing Act (EPF) of 1977

After considerable debate the federal government arranged with the provinces a new cost sharing agreement, the Established Program Financing Act (EPF) of 1977 (Crichton, Hsu and Tsang, 1994, p. 33) under which the annual federal contribution consisted of a cash contribution and a transfer of personal and corporate income tax points (Nestman, 1990). Under this Act, the cash contributions to the provinces were annually escalated by a moving average rate of increase in the nominal gross national product per capita. In 1977–78 there was an additional $20 per capita paid as an incentive to develop extended health care services, such as home care and nursing home care. The formula in the Act ended the link between federal expenditures and actual health care costs incurred by a province. The EPF is a block grant. The federal and provincial governments can, every five years, re-negotiate the terms of the EPF; however, the federal government holds the power and can modify the terms as it sees fit. The constitutional division of powers provides the provinces with the authority to regulate health care, while the federal government "is free to spend its money to influence the way the provinces act" (Government of British Columbia, 1991, p. B-73).

The federal government at first maintained the escalation regime. However, in 1982/83 it changed the rules so that the GNP escalator would apply to the total value of federal transfers such that the cash transfer would be equal to the difference between the total contribution and the value of tax points. Over time the federal government reduced the GNP escalator and in 1990 imposed ceilings on Canada Assistance Plan payments for the top three wealthy provinces – Alberta, British Columbia and Ontario. Also, in this and subsequent years the federal per capita transfers were frozen. The freeze and the ceiling only increased the controversy and tension between the federal and provincial governments. This meant for the EPF that when revenues are increasing from the tax points and the total amount transferred is frozen (cash and tax points), the cash transfer is effectively decreased. In short, the federal government is now "pushing the lid down" on its contributions simply by holding down the rate of growth in total transfers.

Canada Health Act

After a great deal of debate the federal government passed the Canada Health Act in 1984 (Taylor, 1986, p. 206). The Act laid down the basic principles of how health services would be administered with the provinces (Government of Canada, 1984). Five principles were established.

1. *Comprehensiveness* requires coverage of all medically necessary health services provided by hospitals, medical practitioners and dentists.
2. *Universality* requires that 100% of the population be insured for health services under uniform terms and conditions.
3. *Portability* requires that persons moving from one province to another be covered.
4. *Public administration* requires that administration will be by a public non-profit authority and responsible to the provincial government.
5. *Accessibility* requires that there must be reasonable access to insured health services unimpeded either directly or indirectly by charges or other means.

Health services funded by a provincial plan under the Act cannot be funded simultaneously by private insurance companies (Nestman, 1995a). This allows the spreading of the risk of poor health over the entire population and prohibits low risk or high income individuals from opting out and seeking private insurance. The outcome is a one-tier system of health care for those services covered by the Act. Physicians must operate their practices either all-in or all-out for these services. They cannot, as in many other countries, be both in for some patients and out for others. It is technically possible for physicians to fully opt out of a provincial plan and practice on a private basis. However, their patients would not have either private or public insurance and as a result this option has not developed.

The provinces agreed to these five conditions, and as a result effectively ended by 1989 the practice of extra billing by physicians and hospitals for inpatient and outpatient user charges. A province that decides that its hospitals and/or physicians should utilize user charges is penalized by the federal government; the total of such charges are deducted from the funds that are allocated to the province.

Federal and Provincial Health Care Studies

During the 1970s and 1980s the federal and provincial governments embarked on a series of major studies to evaluate the health system. What is remarkable about these studies is that, in general, similar reform language, rationale and recommendations were made. Each of the studies (Nestman, 1995a) concluded that:

(a) the fundamental principles of the Canada health system were sound
(b) the resources devoted to health care services were sufficient to meet the needs
(c) the great health equation was not true, that is, the more that is spent on health services the higher the health status of the population
(d) much of the health care that was provided was ineffective, inappropriate, expensive and not evaluated
(e) the present governance structure in the health care system is antiquated and obsolete
(f) community-based care and regionalization of said should be emphasized

(g) physicians were not the sole deliverers of health care but are a part of a multi-disciplinary team
(h) alternative methods of payment should be considered to the fee-for-service system because of its built-in incentive to increase the volume of services (Hastings and Vayda, 1986).

The reports documented that equalizing access to health care by removing the financial barriers and increasing the quantity of resources available had not equalized health status across the population, and, in order for there to be further improvements in health status there needed to be an improvement in individual life styles. Two notions were agreed upon:

- that the focus of the health system must be on providing and funding only those services that improve health outcomes
- that many of the sources of mortality and morbidity are beyond the capability of the conventional health care system and that more extensive interventions are needed by other areas of the economy.

Despite the effort, very little structural reform took place in the 1970s and 1980s. The existing health care system was still regarded as sustainable. There was some success in moderating expenditure increases by the use of global budgets. Administrative limits on allowed increases compelled hospitals to close beds and to develop more effective and efficient processes for admission, discharge and personnel staffing. Additionally, many joint ventures and private contracting in support services such as food services, laundry, biomedical engineering etc. were undertaken (Hastings and Vayda, 1986).

1990s – A Period of Major Structural Change and Cost Reduction

Health Care Costs and Health System Issues

In 1980 Canada was spending 7.5% of gross domestic product (GDP) on health services and experienced real terms growth of 5% per annum in health care spending throughout the 1980s, twice the rate of growth in GDP. By 1995 health care expenditures accounted for approximately 10% of GDP and 10% of the labour force were employed in health care provision (Blomqvist and Brown, 1994). Federal and provincial government accounts are responsible for 72% of health care spending (Tholl, 1994). From 1975 to 1991 the proportion of health care costs that was financed by the private sector increased from 23.6% of total health care spending to 27.8%. While this increase reflects de-insuring some services, more importantly it reflects the growth of those services in which the private sector has played an increasing role, such as in dental care, long-term care institutions and pharmaceuticals. As increases in public sector spending taper off, the proportion financed by the private sector escalates to fill the gap. This means that those components of health spending that are under direct governmental control are only keeping pace with the growth in Canada's gross domestic product and those sectors that are unregulated and privately financed continue to grow in both real and relative terms.

Additionally, access problems still persist for rural inhabitants, native Canadians, and those who require chronic care, rehabilitation services, home care and some high-tech services. Long waiting lists for certain diagnostic and therapeutic services, such as coronary bypass surgery, have been highly publicized throughout the country (*The Globe and Mail*, 1995). It is now well known and spread by the media that many services are used inappropriately and excessively (Government of Canada, 1994a; Mickleburgh, 1993; Redelmeier and Fuchs, 1993). A strong view has emerged that simply cutting hospital budgets and reducing physicians' fees would not necessarily cure the problems. Canadian health care providers have had many years of budget restraint, but excess patient admissions and inefficient service provision still persist. Cutting global budgets may reduce costs, but cutting as the only measure will not improve efficiency or effectiveness (Brook, 1994). Lastly, health care analysts have become alarmed because medical schools graduated so many physicians by 1989 that the growth rate in the supply of physicians was over three times higher than the general population growth during the 1980s (Rachlis and Kusher, 1994).

Health Care Reforms

By 1993, governments in Canada concluded that the tenure of the existing health care system was no longer viable and a health care affordability crisis existed. Today deficit reduction is a major governmental policy priority and this has increased their resolve to increase the efficiency and effectiveness of the system, and reduce costs. The provinces are in the process of major reform. Some reforms are listed below.

- De-insuring services and using user charges for services that are not covered by the Canada Health Act.
- In many provinces, health services are being reorganized on a regional and community basis.
- Cuts in medical school enrolment – 10% reduction (Government of Canada, 1992a).
- Modifying fee schedules to change practice patterns, instituting caps on individual physician earnings and a fixed budget on total funds available for physician payment and encouraging other payment methods.
- Freezing and cutting public sector wages and salaries.

- Major cuts in acute care bed supply and staff positions. For example, the Province of Saskatchewan closed 52 hospitals in 1993.
- Expansion of prevention, community services, home care, long-term care and a growing concern about outcome measures and practice guidelines.
- Some provinces are increasing co-payments for drugs publicly financed and are removing classes of drugs altogether from formularies.
- Soft caps global budgets, where governments will pick up the over-expenditures are becoming less frequent and most are moving to hard caps where the risk of over-expenditure rests with the provider.

Changes in Federal Transfers to Provinces

As was noted earlier, the federal cash transfer portion of EPF has over the years been decreasing, and as a result the provinces bear more of the risk for increasing health care costs. Today, health care analysts are concerned that federal cash transfers will eventually disappear. Because the cash transfer is the only amount shown as a federal contribution in provincial accounts, its disappearance will have the same effect as the termination of the entire EPF agreement and long before this point is reached the declining federal EPF cash transfers will call into question the legitimacy of a continuing federal management role in health care (Government of British Columbia, 1991, p. B-73). The key issue is how the federal government will be able to enforce the principles of the Canada Health Act, when the incentive of withdrawing funds for provincial non-compliance will disappear.

When the federal government delivered its 1995 budget it folded those programs funded under EPF and those under the Canada Assistance Plan into a new transfer regime called the Canada Social Transfer (CST) (Government of Canada, 1995). This provides more predictability for the federal government because the CST is a block grant in that the amount of funds that will be transferred will not be determined by provincial spending decisions. The Canada Assistance Plan is thus the last major cost-sharing program to come under a fixed budget system. The new transfer will be provided through a similar cash payment and tax point system as the EPF. The provinces will have a wider scope of discretion to allocate resources to health, post secondary education and social services. They will receive the same amount of funds from the federal government in 1995–96 as they received in 1994–95. However, payments will be cut – 2.5 billion for 1996–97 and 4.5 billion for 1997–98. In addition, the federal government has indicated it will uphold the standards of the Canada Health Act and will "invite all provincial governments to work together on developing, through mutual consent, a set of shared principles and objectives that would underlie the new transfer".

ANALYSIS AND OBSERVATIONS

Physicians and Acute Care Hospitals

In the 1980s provincial governments used global budgets as the main cost control method. Very little direct effort took place to control hospital worker incomes or bed capacity. Hospital unions and local communities would have objected in the strongest possible terms. Negotiated fixed budgets provided an invisible lid on costs. Painful policy decisions were avoided. Hospital boards were used as a buffer between governments and workers and/or communities. In the 1970s and 1980s global budgets did have an impact on restricting the number of labour inputs and beds that were kept open. In Ontario from 1971 to 1981 the total annual paid hours per 1000 population fell from 26 000 to 24 900 and paid hours per patient-day increased only 0.6% compared to the growth of almost 1% per year in the prior decade (Lomas and Barer, 1986). The number of beds per 1000 and the number of hospital beds per physician for this same period dropped from 6.40 to 6.23 and 4.22 to 3.35 respectively. It is interesting to note that the population per active physician ratio during this same period dropped from 659 to 538 (Barer and Evans, 1986). Thus, the situation in most provinces was that the bed supply and hospital budgets were being restricted, while the supply of physicians continued to increase.

By squeezing global budgets, provincial governments were able to tighten the relationships between physicians and bed capacity, and physicians and health personnel employed in hospitals. Clearly provincial governments chose the policy that had an impact on moderating the increase in hospital costs. However, it did at the same time distort the health system by not dealing with the real issue – the increase in the number of physicians at a faster rate than population growth. In the mid 1980s provincial governments realized that their indirect control methods were not sufficient. It was the 1990s before major direction occurred. Today, provincial governments have moved beyond blunt instruments of control, to a massive restructuring of the system. The pay offs to the health system may be high, but, the political costs may be even higher.

Many health critics feel that the increasing yearly number of physicians has been a major contributor to the rise in health care costs (Evans, 1992). Since 1989 the population to physician ratio has stabilized at 514 in 1992 (Tholl, 1994). Some feel that as the 10% cut in medical school enrolment comes into effect, the growth in the number of physicians should drop below population growth for the first time in Canadian history (Barer and Stoddart, 1991). It is, however, the increase in the intensity of services per patient provided by physicians and the fee-for-service basis of payment that have been the major factors contributing to increasing health care costs (Government of Canada, 1992b). There are signs that for the period from 1987 to 1991, the actual utilization, more specifically the rate of increase in service intensity, has been moderating and in some provinces has been dropping. A change in practice style and a reduction in the number of acute care beds contributed to this moderating trend. However, the commencement of spending caps on individual physician earnings and firm, negotiated, spending envelopes for physician services are the major factors that have contributed to slowing down the rate of service intensity.

If physicians individually or in total exceed the limit established, they absorb all of the over expenditure themselves (hard caps) or in some cases a negotiated portion of it (soft caps). As a result, the incentive to increase services that the fee-for-service basis of payment usually provides is lessened. A concern is now emerging about what impact the de-insuring of services will have on the growth of services that are provided by physicians on a private basis (cost shifting) where the effective control systems are not in place. While the recent data that are available indicate that physician services that are financed from private sources have actually gone down from 4.4% to 4.0% from 1987 to 1991, one could anticipate in the future some enlargement in the private sector share because of the wave of de-insuring that took place after 1991.

Due to the tightening of global budgets and the continuing trend to downplay acute care programs, hospitals have continued to withdraw beds from service (cost reduction) and/or convert short-stay beds to long-term beds (cost shifting). Total hospital costs as a percentage of total health care spending have remained relatively stable at 39% from 1987 to 1991. Unfortunately, information is not yet available that would indicate the impact of more recent hospital budgetary actions and bed reductions. There is, however, a slight rise from 11% to 11.5% in the private sector share of hospital costs during the same period, which probably represents a small degree of cost shifting. There are, however, costs that physicians generate in hospitals, such as the cost incurred in tests and therapies ordered, etc., that because of data limitations cannot be singled out for analysis. There is no doubt that these costs are one area of focus in the recent efforts to curtail the number of physicians in the system.

Canadian hospitals could change the method of reimbursement from global budgets to a prospective payment case-base payment (CBP). A few provinces are using this method and they have been influenced by the experience in the USA (Kahn et al., 1990). Some initial evaluations have shown positive aspects to improving efficiency in the hospital system (Schurman and Lynch, 1994). Soderstrom (1984), however, notes that the adoption of a CBP system would probably not significantly improve the performance of hospitals in Canada and would increase the administrative costs of the system. Governments do already have an effective means to control costs – fixed global budgets. Soderstrom notes further that financial incentives inherent in CBP's may not be superior to those of global budgets. In many respects, the incentives are the same: hospitals gain by minimizing their costs for producing any given service. Where they differ each has positive and negative effects. CBP provides a stronger incentive to reduce length of stay because doing so minimizes costs per case. However, it also provides an incentive to admit more patients; a hospital can gain financially by admitting all possible cases whose costs would be less than their CBP fee, whereas global budgets have the opposite incentive.

CBP also has an advantage in that it provides a means to distribute funds on an equitable basis among hospitals because it provides a means to adjust for case mix differences. However, recent experience in Alberta indicates that CBPs do not assure an equitable distribution of funds because they can be manipulated to favour some hospitals at the expense of others (Jacobs, Bay and Hall, 1993; Plain, 1993). It is still too early to provide stronger conclusions as more time is needed to properly evaluate the use of CBP.

Unfortunately, there has been very little evaluation of the impact of budget constraints on quality and the utilization of health services, even though fixed budget cost control efforts have been the cornerstone of health care policy for many years. The funding of health service budget methods and constraints have been based in large part on casual and anecdotal evidence. Detsky et al. (1990) have observed: "It may be in the interest of all parties, including government, institutions, clinicians, to have the facts remain unknown; in that way rhetoric and orchestrated outrage can be freely used without fear of being proved wrong."

There are, however, a number of observations that could be made concerning global budgeting systems. First, historically the incentive of global budgets has been considerably weakened in many provinces by the inclination of governments to cover deficits at the year end. However, provincial governments in the 1990s have become considerably more firm in not picking up spending over budget. Second, it is impossible to isolate the impact of global budget systems to cost containment objectives alone. Third, when most physicians are paid on a fee-for-service basis and most other health care professionals are paid on a salary basis these payment systems have major implications for resource allocation. It follows in turn that the level of resources consumed in the patient care process is determined, in large part, by individual physicians. As a result, health service budgets are set implicitly, not explicitly, and this makes the changing of the allocation of health care resources very difficult to achieve, particularly if the changes are being attempted from those outside the medical profession. Fourth, physicians, as a group, are not responsible in a line management sense to administrators, and it is therefore misleading to think of the hospital as a well defined economic entity that is under the control of a single central source, such as the chief executive officer. Physicians are not directly held responsible for costs incurred in the hospital. This situation has a disabling and tension provoking impact on the health system because the financial payment systems are in conflict in terms of their impact on hospital operations (Nestman, 1989) (see Table 9.1).

Table 9.1 Financial payment systems

Hospital cost influencing variables costs	Volume	Admissions	Input
Physician fee for service	+	+	+
Hospital global budget	–	–	–

The fee for service system encourages increases in volume, both in and out of the hospital. Hospitals, on the other hand, are confronted with global budgeting systems that tend to cap the level of health expenditures and thereby apply pressure to reduce volume. While admission and input costs will have a tendency to increase under the fee-for-service system, the global budget based system will encourage administrators to reduce admission and input costs because of the restraint of a fixed budget. These then, are powerful forces, both pushing and pulling in opposite directions.

Regionalization

One of the key questions of the reform process is whether the new regional governance structure can improve the performance of the health care system. The governance of the new agencies varies from province to province; some members of the new agencies will be either elected, appointed, professional or non-professional or a mix of the above. Some provinces have initially restricted the scope of services administrated by regional agencies to only acute hospitals (Government of New Brunswick, 1992), others have included a wide spectre of services on the continuum of care (Government of Saskatchewan, 1993). So far, physician and pharmaceutical services have been excluded from the regional agencies. Community health boards assume responsibility for the management of and allocation of budgets to health service organizations and receive their funding from the regional agencies. In those provinces that have not or intend not to set up community boards the regional boards assume the functions of the community health boards. The ministry of health continues to provide some specialized services, distributes funding envelopes to the regions, develops provincial polices and plans, and sets standards, evaluates programs and monitors outcomes.

Some provinces (British Columbia, Saskatchewan, and Nova Scotia) have devolved authority from the Ministry of Health to local bodies (regional and community councils) which have a clear legal mandate and considerable degree of autonomy (Hurley, Lomas and Vandna, 1994). In these provinces the ministry of health has not prescribed a plan for change that regional and community agencies are to execute. Instead the specific changes that will occur will be decided and implemented by the local bodies. In contrast, New Brunswick and Quebec have been delegated certain managerial functions and plans for them to execute, and they will be accountable to the Ministry of Health. In Quebec the Ministry of Health has retained a great deal of authority over regional boards. In contrast, in Manitoba no regional boards have been set up and the power to change the health care system has remained in the Ministry of Health (Government of Manitoba, 1993). Ontario has set up regional councils; however, in general their powers have been advisory in nature only. The reforms in Saskatchewan, Nova Scotia and British Columbia, and to a lesser extent in Quebec, are designed to alter the balance of power between citizens and government and efforts so far have a strong populist tone. In contrast, New Brunswick the reforms are aimed at altering the balance of power within a new governance hierarchy structure.

It is too early to assess and provide an effective evaluation of the impact of these changes. Disputes will probably continue about the powers that will be relinquished to the new regional and community agencies. However, it is very clear that accountability for health services as designated by the constitution continues to rest with the provincial government. The provincial government will still have the ultimate power of the public purse and it will as a result be unable to elude ultimate responsibility and accountability for the use of resources. In fact, the disbanding of hospital boards and the creating of regional boards that would closely oversee (in some cases actually own) the operations of hospitals can be regarded as a significant shift in greater govern- mental control. Governments in Canada have adhered to the notion that a minister of the crown should maintain control of the public purse because he or she is ultimately responsible to parliament for the use of those resources. In the province of New Brunswick accountability for budget planning and control runs directly to the Ministry of Health, which is then accountable to the people through the elected government. Ministerial authority has remained basically intact. It is noteworthy that little has been done within the ministry to increase accountability to the people. In contrast, in the provinces of Nova Scotia, Saskatchewan and British Columbia accountability is jointly shared with the provincial government and the residents in each region. Accountability to the provincial government is maintained by the normal budget process. Accountability to the people is maintained by the election and/or appointment process and by open board and information meetings.

All the provinces that have created regional or community councils have adopted fixed budget funding envelopes. Within the restraints of the fixed budget provided by a province the regional or community councils have the power to allocate funds. There are a number of positive points one could make regarding the effectiveness of regional fixed budgets. A fixed budget will make it easier to redistribute resources from areas of over capacity to areas in need as well as to foster transitional efforts to move labour resources from acute care to community services. If a community is involved in the decision to make a change it is more likely to accept the change. In that sense supply-side management would be shifted from the department of health to the local area which, in turn, would be responsible for managing services for a smaller number of people.

There are, however, some potential difficulties with this approach. Canadian and international experience has provided strong evidence that cost control is fostered when sole payer and fixed budget methods are in operation (Nestman, 1995a). The benefits are the ability to consolidate administrative and planning functions and payment systems. As a result, the health care system has been able to achieve administrative, economic and informational economies of scale, that more diverse and pluralistic health care systems have not been able to achieve. The provinces seem to have realized this; they have adopted fixed regional funding envelopes. However, fixed budgets and the regional boards provide a local mechanism, that has political legitimacy with constituents behind it, that could pressure for more resources from the central authority and could cast blame for underfunding and/or service inadequacies on the central authority. In addition, the local body may attempt to have access to revenues from local governments (municipal-property taxes). Instead of a sole payer system, a dual (province-region) or multiple (many regions) payment system might ensue. This could result eventually in service inequalities developing within various regions in a province. The questions to consider here are, what burden of health care costs is a province willing to bear for administering its health system (when we know that multi-payer systems cost more to administer) and how much control is the province willing to lose for the sake of diversity and pluralism (when we know that sole payer systems have the strongest control)?

One of the important basic concepts currently being encouraged is to achieve a better balanced health care system so that a more effective and efficient continuum of care can operate. The legislative, administrative and financial framework for health services should facilitate the achievement of this objective. When health insurance was developed in Canada, the medical care and hospital programs were financed, at both the federal and provincial level, under separate legal and financial frameworks. In effect, provincial delivery systems mirrored the federal system and hence provided a strong accountability mechanism so that expenditures could be tracked and audited to comply with the needs of the 50–50 cost-sharing arrangements. Additionally, public health and mental health services, which were not covered by federal legislation, were administrated and financed by the provinces themselves under still yet separate legal and financial frameworks. When federal block funding under the EPF came into existence in 1977, provincial governments still continued to isolate and separate all four delivery systems – legally, financially and administratively. Fixed budgets helped to cement this rigidity because when attempts were made to transfer patients or develop systems that would facilitate co-ordination across the continuum of care, the financial distortions inherent in fixed budgets did not allow the funds to follow the patients throughout the care process. The result was that health legislation and block funding helped confirm the isolation of the various sectors of the health services system (Nestman, 1970).

As provincial regionalization programs take hold, ministries of health and regional authorities must make sure that what happened at the federal and provincial levels does not repeat itself at the regional and community levels. So far it does not appear promising that physician and pharmaceutical services have not been devolved to the regional levels, and that separate administrative, financial and legal structures persist. For the most part the provinces have treated these sectors as separate policy initiatives under their own preview. What is now happening is that budget planning, control and management of significant interrelated parts of the health system are administrated by two different levels.

A significant issue relates to the size of some of the regions. Most Canadian provinces are small, in terms of population. Creating further divisions may be ill-advised. Indeed there is strong argument that interprovincial centres of specialization and co-operation mechanisms would be more effective and efficient. Prior to the introduction of regional and community agencies, budget planning and decision making took place at the provincial level and the expertise to administer the system was developed and maintained at that level. Decentralization of significant budget powers will require that planning and budget expertise be developed and maintained for that level. The question can be asked whether this level of expertise development and maintenance will lead to a duplication of effort and an increase in administrative costs which are not warranted considering the size of regional and community agencies. Further, could not this task be more effectively and efficiently done at the provincial level? It is apparent that the process of regionalization will have to evolve further before we know if these excesses become evident and whether corrective action will be necessary.

While many provincial governments have set up regional bodies to fund and administer regional health systems, this is often an attempt to ask locally based bodies to make some of the tough decisions that provincial governments made or did not make in the past. The implications of this change have not been thoroughly thought out and planned. For example, if a province has ten regions, will ten funding and management control systems evolve? What will the needs be for uniformity and provincial standard setting? Funds that are raised by the provinces and federal governments to provide services under the Canada Health Act are provided to providers with the understanding that all citizens are entitled to the same basic services. Will the magnitude of local decision-making powers be such that a community can decide to cut services that are regarded part of the basic package, and if so what will the appeal procedures be? Will provinces feel the need to legislate the principles of the Canada Health Act and play the role of standard setter and controller that the federal government now plays with the provinces? Will this mean that as regionalization evolves the provinces will be required to introduce strong provincial controls, policies and guidelines? If so, how extensive will this recentralization be?

One application of regionalization that has received a great deal of attention is the population-based funding method (Birch and Chambers, 1993a, b; Eyles and Birch, 1993). The amount of funds that a regional agency receives is determined in advance and is based on the number of residents within the boundaries of the authority (Government of British Columbia, 1993) and on an equal per-capita payment that is then adjusted using mortality rates and socio-demographic factors etc. In a fixed budget all care received by residents must be paid out of the budget even if the care a resident receives is outside the boundaries of the authority. This advantage overcomes some of the problems that currently exist for funding based on per capita health care spending by geographical location. For example, if budget determination for different regions in a province is based on current per capita costs, then those regions that have a higher proportion of urban population will receive a higher budget than those that have a higher proportion of rural population because that is where the high cost resources are located. Under a population-based funding approach this inequality would disappear because funding would follow the population according to their health needs, and hence, a more equitable distribution of resources would ensue.

Pharmaceuticals

There has been a major increase in costs for pharmaceuticals funded by both the private and public sectors. Pharmaceutical costs as a percentage of total health care spending rose from 8.9% in 1980 to 13.8% in 1991 at an average annual increase of over 10% (Tholl, 1994). In 1991, 72.4% of pharmaceutical costs were financed by the private sector compared to 75.9% in 1987. This trend will change because of government action in recent years to increase co-payments, and de-insure and remove drugs from formularies. In addition, there are reductions in the coverage of drugs financed by the private sector through employers. As a result, more and more Canadians will face out of pocket costs for drugs.

Capital Expenditures

The authority vested by legal statutes gives the various provincial departments of health the power to approve or disapprove almost all capital expenditures. This power has been effectively exercised, which has contributed to the gradual annual decline in the percentage of total capital spending to total health care spending from 1975 (5.0%) to 1991 (3.6%) even though the physical infrastructure is 30 to 50 years old (Tholl, 1994). The proportion of capital spending that is financed by the public sector has fluctuated sharply over the years from 70% to 58% and in most cases is a result of provincial election cycles. Provincial health care systems do acquire most of the state-of-the-art technology. However, the most advanced and costly equipment tends to be in tertiary health service organizations and does not proliferate elsewhere in the health care system. Evans (1992) notes that "the availability per capita of such equipment tends to be lower than in countries such as United States, Germany, or Japan".

There are many policy questions that are outstanding in Canada in regard to capital spending. For example, will the general shift from institutional care to community care decrease the need for large capital infusions? If fixed budgeting methods are to remain in place, and community care requires more resources, will the funds be taken from, and at the expense of, capital expenditures? As physical plants continue to age, how long can capital expenditures be postponed? Because Canada's health care system has been acute care focused and provincial governments have had control over hospital budgets, governments were effectively able to control where and to what extent expensive technology and capital expenditures could be introduced and where they should be located. As decision making is dispersed to regional agencies, will strong control be weakened and made less effective?

While control of capacity through budgets has been effective, capacity constraints do not always work and the constraints can cause their own problems (Deber and Leat, 1987). For example, when physicians continue to order services for their patients and capacity restraints delay the services from being provided, frustration and the creditability of the system will be questioned. There is constant pressure from health professionals, patients and family members (much of it coming from media coverage in the USA) to purchase high-cost technologies. While budget methods have been effective in controlling capital costs, better methods are needed to aid adoption decisions (such as effective technology assessment). Funding mechanisms that encourage a higher adoption of new technologies and more capital expenditures exist elsewhere in the world, particularly in the USA, and these have very important spillover effects for Canada (Feeny, 1994). Considering that the Canadian health care system must function in a North American environment, it is quite remarkable how effective cost control has been for capital expenditures.

Budget Control and Community Health Services

Economic difficulties may provide opportunities for creative restructuring; there is, however, some danger that panic and short-term benefit thinking could rule the health care policy agenda. First, the shift from acute care to community-based services represents a change from proven and effective global budgets in the acute sector to a sector where budget control and effectiveness have not been emphasized and proven. As a result new and/or improved effective management control systems will have to be developed. We cannot assume the incentives that are built into fixed acute care funding methods are also appropriate for community-based services when the need will be to expand services.

Second, labour costs in the community sector are lower than in the acute care sector. There will undoubtedly be efforts to equalize this situation and these efforts will be resisted and/or delayed by governments. However, it is clear that once labour costs are equalized the cost-effectiveness of community services over acute services will narrow and it is important that effective evaluations take place.

Third, one of the strategies of moving services to the community sector is to shift tasks from high-cost acute care to low-paid or unpaid volunteers and families. However, it is unclear what capacity community members have to assume this responsibility. For example, will family members have the time to assist those in need in their homes (Nestman, 1995b)?

Last, the pressure of fiscal reality may make the process of gradual and thoughtful reform difficult. Some provinces such as Alberta are attempting rapid and massive cuts. There is the risk that costs will be cut in the acute care sector and shifted too quickly to the community sector and as a result the quality of services will not be emphasized. Indeed many commentators have indicated that it is proving difficult to relocate, in the community, some of the resources cut in acute care (Rachlis and Kusher, 1994). In other words, funds are leaving the health system to fight the deficit. This could impair the credibility of the new reformed health care system and could eventually erode its support.

Impact of Fixed Budgets on System Reforms and Improvement in the Quality of Care

It has also been customary that when the provincial government cuts budgets, the cuts are made on a uniform basis across all hospitals. As a result, hospitals have been undertaking rationalization schemes, such as closing beds. However, it is well known that there is a relationship between volume of services performed and quality of outcomes (Nestman, 1989). If health services are to be provided at an acceptable quality level there is a minimum level of activity that must be provided. Problems arise when hospitals, after the cuts, still perform the same services at lower volumes. It is here where quality of care suffers because it is sometimes difficult when volumes are low to justify the critical mass of resources needed to perform the particular service. It is important that provincial governments or regional authorities restrain hospitals from offering certain services when minimum volumes may not be attained. In addition, closing beds may not be effective – operating a 700-bed hospital with 500 beds is more expensive than operating a 500-bed hospital. Maintaining a large number of hospitals operating below capacity has both quality and efficiency implications.

The Manitoba Centre for Health Policy and Evaluation (Government of Manitoba, 1993) noted that the major weakness of the global budget system is that it does not tie hospital spending to any systematic evaluation of effectiveness of the services being provided or the effectiveness of the care measured by health outcomes. Fixed budgeting systems are oriented towards inputs rather than outcomes which makes other legitimate objectives such as encouraging resource substitution and program alternatives particularly difficult to implement (Nestman, 1989).

The last six years have seen many efforts to improve the efficiency, effectiveness and efficacy at the micro level, such as quality improvement (Schurman and Lynch, 1995), clinical practice guidelines (Linton and Peachey, 1990), outcome measurement and technology assessment. Whereas prior efforts for improvement concentrated on administrative and clinical support changes, now the emphasis is on those processes that improve the direct patient process itself and where professional resistance will be highest. In addition the lead time for success is long and complex and not within the framework of political agendas. Many of these initiatives would not have moved forward without those who are employed in the health system realizing that there were only limited resources (fixed budgets) available to be used in the health system, and that if improvements were to take place resources for the initiatives would have to come from within existing budgets. However, there is strong opinion that incentives are needed to facilitate these initiatives, and barriers and obstacles must be removed. An understanding of potential and actual distortions associated with financing mechanisms is critical in the current context of attempting to promote new initiatives. In response, the federal government is providing research funds to examine how financial incentives can be improved so that effectiveness, efficiency and efficacy forces can be improved within the system (Government of Canada, 1994b).

The Changing Role of the Federal and Provincial Governments

In our federal-provincial mechanism there are three concepts that are dear to Canadian hearts: accountability, redistribution, and provincial autonomy (Nestman, 1970). Funds should be spent in such a way that governments can account to taxpayers through parliament for the efficiency and effectiveness of expenditures. Also, fiscal redistribution policies must ensure that there is a reasonable uniform level of public services across the country. Furthermore, programs should be provided in a way that ensures local requirements are taken into account. This is difficult. For example, the block funding EPF arrangement emphasizes redistribution and provincial autonomy. However, it leaves the federal government unable to account to parliament for the effective use of funds sent to the provinces. Similarly, the federal government's Canada Health Act mandates that services must be delivered on uniform terms and conditions across the country. This emphasizes redistribution and accountability but is contrary to provincial autonomy. Canadian federal-provincial relations will always create a sense of drama and uncertainty because these diametrically opposing concepts are difficult to balance.

The federal initiative of creating the Canadian Social Transfer, and of pursuing with the provinces the development of new principles to administer these programs will result in a major redefinition of the methods by which health and social services will be administered. It is too early in this process to be able to offer some definite conclusions about what the outcomes will be. However, a few observations will be provided. This amalgamation will require a scope of co-ordination and management that will be far more complex and broad in scope than has been attempted before and it will be undertaken in a climate where funding is decreasing. There is very little agreement on what the minimum requirements are for health promotion, community and social services and income maintenance programs and their relationship to the health system. The national health care insurance program has been Canada's most popular social program, whereas the other programs, on balance, have not enjoyed the same level of support. These programs are more vulnerable and their prospects for erosion and fragmentation within the new and emerging decentralized system appear to be increasing. Political maturity and balanced judgement from elected officials and administrators will be required.

Canadians, by and large, have accepted a strong role for governments in the development and reform of the Canadian health system. However, the role of federal and provincial governments as standard setters and controllers must be carefully considered, particularly if they wish to provide a basic package of health services on uniform conditions across the nation to all their people. Canadians and their governments are facing a major decision regarding their health insurance system. The Federal government must decide if it is going to maintain one national health insurance system or develop ten distinct and not necessarily similar health insurance systems. The provinces have a similar decision to make. Regions have different fiscal capacities. From a national perspective, the greatest threat to the five principles of the Canada Health Act appears to be the continuing decline in federal contributions for health care. If present trends continue, the credibility of the federal government's management role in health care will be challenged. From a provincial perspective a major risk will be the move from a central control system to a multiple control system (regional and community). Provincial governments will find it more difficult to maintain a strong position to control costs and approve programs.

Canada has selected to finance first-dollar coverage for services required under the Canada Health Act. There are very few issues that provoke as strong emotional reactions and debate in Canada as the issue of user charges (Barer et al., 1994; Evans et al., 1994; Stoddart, Barer and Evans, 1994). Some provincial governments have, from time to time, very forcefully expressed a desire to implement user fees. However, even though user charges have been used for many years for pharmaceuticals and long-term care services, their use for hospital and physician services is broadly regarded as incompatible with Canadian values. The federal government has indicated that it will continue to enforce this provision (Marleau, 1995). In addition, because private insurance cannot operate and provide insurance for the services funded under the Act, and because most health care costs are funded under the Act, there is a system-wide control mechanism that can be used by governments to control total health care costs.

There are, however, many services that are now being de-insured wholly or in part. In a sense these services have escaped the controls of the fixed budget system and are now provided by the private sector. For those services that provide a positive health care outcome and are now being paid for by patients or private insurance the end result is cost shifting rather than cost containment. For services that are not being provided to patients because the patients cannot incur the costs on their own, the objectives of the Canadian health care insurance system will have been defeated.

The real question from a national and provincial point of view, in terms of setting national and provincial priorities, is how much their jurisdictions spend on health services in total, both private and public, as compared to other priorities such as education, industry, environment etc. For global budgets to be effective as a cost-control mechanism, provincial governments must be able to control all the financial flows within the public global budget system as well as the financial resources that come into the system from the private sector. To do this, information systems must be able to monitor both public and private flows of resources and provide the pertinent information to decision makers. However, as provincial health care systems become more diverse and complex, the information requirements for decision making by provincial governments, regional and community agencies will increase and so will the costs of maintaining such an elaborate information system.

When health care providers increase volume under hard capped fixed budgets they in effect receive a reduced price per unit of service. However, prices cannot be scaled down indefinitely. When the total health care budget is equal to price times quantity, both components become a focus for cost control targets. The option of using pricing systems has been severely curtailed as a vehicle to allocate resources in Canada. This leaves the focus of downward pressure on volume and intensity. Global budget review systems have relied on negotiation and regulation to determine funding levels and in this context the issue is how health care expenditures can be more competently managed to achieve efficiency, effectiveness and efficacy. For provincial policy makers the only way they may be able to control health care expenditures produced by providers is by being more direct with incentives and regulations.

Accountability requires that providers be held accountable for the efficient use of public resources that they deliver. The issue then arises as to what level of the system does the government play a stronger role – provincial, regional, community, or the provider level? For example, if community boards use their budget power to contract physicians and other health care providers, the distinction between a payer and the provider is blurred – a governmental HMO. The significance of this shift from provincial governments as a payer of health services to health service manager is immense. Indeed, does this mean that as resource flows to the health care sector decrease, a sole payer, utilizing a fixed budget, will be ultimately compelled or obliged, by its requirement for cost reduction and efficiency, to intercede in the health system? A pillar of the Canadian health system has been the compromise between professional autonomy and public fiscal responsibility and accountability.

An objective of reform process was to change the health care system so that results were acceptable to both the providers and consumers alike. One of the advantages of fixed budgets as a cost containment measure is that they introduce a sense of discipline into the system without getting the government (or a regional or community agency) involved in micro management. This balancing act is delicate and there are signs that the emphasis of governmental efforts in cost control and reduction may change and become more direct. Some governments are considering more actively directing practices and delivery of services by stronger utilization controls, such as providing incentives in agreements with physicians to enforce protocols to restrict utilization, to move physicians to rural areas, to deny or restrict admitting privileges, etc. Other examples are the health services organizations (Ontario) (Adelson and Birch, 1993) or community health centres (Quebec and Saskatchewan) (Rachlis and Kusher, 1994) that were developed with the expectation that they would reduce costs and improve quality by using other payment systems such as salary and capitation.

One option that is actively being pursued is changing who can make patient care decisions. Some provinces are now broadening the decision-making scope of some health care personnel, e.g. expanding the role of nursing and including new health care personnel, such as midwives and nurse practitioners. Many of these reforms are being undertaken with a significant degree of resistance from the medical profession. It is too early to tell if these interventions will continue and grow in scope. If they do, the nature of the health care system will change fundamentally. They do have one common component; these approaches bring the management of health care resources closer to the actual patient decision making than occurred in the past. At the centre of this issue is the relationship of the provincial government as a manager to the health care professions who are responsible for making patient care decisions.

SUMMARY AND CONCLUSIONS

It is clear that the Canadian health system is now in an era of significant structural change. The dramatic reforms that are initiated are dominated by the concepts of decentralization, health promotion and community services. The national health care system rests on the changing authority of federal government, which has exerted its influence by providing funding, that is decreasing in real terms, and by exercising its power to withdraw funds under the Canada Health Act if provinces do not comply with the principles of the Act. The 1995 federal budget will embark the country on a new process to redefine the national health care system for the twenty-first century.

The Canadian health insurance system is facing a major decision. Will there be one national health insurance system or ten distinct and not necessarily similar health insurance systems? Provinces in Canada have different fiscal capacities. From both a national and provincial perspective, the greatest threat to the five principles of the Canada Health Act appears to be the continuing decline in federal contribution for health care. If present trends continue, the credibility of the federal government's management role in health care will be challenged. The federal government must assume the major responsibility because it is the only government which can establish national standards, and enforce these standards by eliminating its financial support. Provinces must similarly decide how they are going to allow their health systems to evolve so that the popular principles of The Canada Act can be applied province-wide. The pressure of adjusting the present health care system to one in which community-based services are given a higher priority means reallocating resources. The difficulty in making some of these changes is that a significant reallocation of resources must take place. Health care expenditures are also people's incomes. Resistance has emerged. Yet a consensus has formed in Canada that the health care system needs to be improved so that only appropriate care is delivered at a high quality. Canadian health care policy has over the years displayed one consistent tendency – that is, an explicit commitment to inspire health care organizations to achieve quality, increase access, improvement equity and contain costs.

When the health care insurance program was first developed in Canada the nature of Canadian society was very different from that of today. There were shortages of hospitals, physicians and other health personnel. The payment mechanism at the national level was a cost-sharing system which allowed provinces to determine the level of expenditures. At the provincial level, physicians were paid on a fee-for service basis and hospitals were on a line-by-line budget at first, then a global budget. These budgeting systems as applied in Canada up to the mid 1980s were cost-based reimbursement systems. The health system was in a growth and expansion phase and as a result these systems promoted volume and encouraged productivity.

In the mid 1970s the tempo began to change when the federal government enacted a block funding payment mechanism to pay for their share of health services to the provinces. This was a reflection of the federal government's concern with limiting its risk in terms of total health care funding, its anxiety with fiscal issues and its realization that increasing health care costs beyond the existing levels may not provide positive health care outcomes. The provinces had similar concerns, and one other. Federal action was resulting in the risk for increasing health care costs being shifted to their jurisdictions. The provinces responded by capping physician expenditures and applying hard caps to global budgets. In effect, as the situation in the country and health care system changed so did the funding methods throughout the system. In a sense, as the needs at the macro level required fixed budgets, a cascading effect resulted. Fixed budgets at the meso and micro levels developed. What a health system needs, is flexibility to alter methods to meet changing needs and circumstances in the various sectors of the health system. For example, funding methods that encourage volume growth in the community care sector may be appropriate for a time to foster expansion.

Fixed budgets as a funding system are now under pressure because they have resulted in some distorting effects and they have not encouraged promotion of appropriate care. What is needed now is a payment system that will provide the incentive to encourage the pathways for the reforms that are desired. Health care policy makers are being pressed to actively manage health resource use. A number of important questions are pertinent here. How will the Canadian health care system, as it moves into the twenty-first century, develop a cost conscious decision-making climate that will be congruent with the principles of a universal publicly sponsored health care system? Since policy makers may be unable to make health care practice improvements through global budget strategies alone, will incentives (financial or others) and regulations be necessary and will this change the authority and role of government as a payer, planner, direct provider at the provincial, regional and community level?

Canada may be entering an era when the downward pressure on prices and cost shifting will eventually be replaced by a more direct public management of the system. The overall lesson, however, that stands out in the Canadian context is that the use of fixed budgets and regulation of supply, rather than demand, can provide a vehicle to control hospital costs. However, fixed budgets and regulation of supply, by themselves, will not address many fundamental issues for the future, and may actually be part of the problem.

When the cost of health services is covered by a sole insurance provider under specific terms and conditions, the insurer can have a great deal of influence on the restrictions governing the level of funding. However from the Canadian experience it appears that governmental cost control effectiveness depends on the extent of health expenditures funded. The locus of control in a sole payer system is the government. While the power of governments in Canada to control costs still remains in place and intact it also follows that as the proportion of health care funded by the private sector increases, government ability to control diminishes commensurately. Canada's health care system is in the midst of an affordability and cost shifting crisis which should be clearly distinguished from a cost containment crisis. The policy of cost shifting to the private sector is a policy that will result in short-term gain and long-term pain. Governments in Canada should deal with this sooner rather than later – the pain will only become more acute as time passes.

An individual country's health care system is a reflection of its historical, cultural and economic values. Health care providers are powerfully influenced by the specifics of their national and local funding systems. Countries usually develop their own distinct practices for paying and organizing their services. The concept of fixed budgets has contributed in a major way to the development of the Canadian health care system. The question one could ask is why has Canada made this choice to control health care costs and why has it been on the whole comfortable with this choice? Certainly it does meet the government's need (and for that matter society's need) to establish control and predictability in the level of health care costs. However, it does reflect Canadians' acceptance of central authority and the importance they have bestowed on public financing of health services. Global budgets are an encompassing, consistent, firm and precise fiscal mechanism and their use and acceptance as an instrument of public policy reflects the social culture of the country. The role of government as noted in the Canadian constitution is to provide peace, "order and good government". Hence global budgets have been accepted as an important tool. It is acceptable in Canada for governments to be involved in regulation. Once the principles are agreed upon, professional health service administrators will settle, through negotiations, resource allocation issues. Fixed budgets will continue to play a major role in shaping the Canadian health care system, and the changes that are unfolding will in turn impact on how fixed budgets will be applied. Ultimately, the use of fixed budgets and their application will be judged on how their use will be consistent with a universal, publicly financed health care system and how they maintain the principles of the Canada Health Act.

REFERENCES

Adelson, J. and Birch, S. (1993). Alternative funding and delivery models: practice and prospects in Ontario. *Journal of Ambulatory Care Management*, **16**, 3, 19–29.

Barer, M.L. and Evans, R. (1986). Riding north on south-bound horse? Expenditures, prices, utilization and incomes in the Canadian health care system. In: R.G. Evans and G.L. Stoddart (Eds), *Medicare at Maturity: Achievements, Lessons and Challenges*. Calgary: The University of Calgary Press.

Barer, M.L. and Stoddart, G.L. (1991). Toward integrated medical resources policies for Canada. Report prepared for the Federal/ Provincial/ Territorial Conference of Deputy Ministers of Health Centre for Health Services and Policy Research. Vancouver.

Barer, M.L., Bhatia, V., Stoddart, G.L. and Evans, R.G. (1994). The remarkable history of the participation, positions, and rationales of Canadian interest groups in the debate over direct patient participation in health care financing. Paper presented at Premier's Council on Health, Well-being and Social Justice. Toronto: Government of Ontario.

Birch, S. and Chambers, S. (1993a). A need-based approach to resource allocation in health care. *Canadian Public Policy*, **19**, 1, 68–85.

Birch, S. and Chambers, S. (1993b). To each according to need: a community-based approach to allocating health care resources. *Canadian Medical Association Journal*, **149**, 607–12.

Blomqvist, A. and Brown, D.M. (1994). *Limits to Care: Reforming Canada's Health System in an Age of Restraint.* Ottawa: C.D. Howe Institute.

Brook, R.H. (1994). Appropriateness: the next frontier. *British Medical Journal*, **308**, 218–19.

Crichton, A., Hsu, D. and Tsang, S. (1994). *Canada's Health System: Its Funding and Organization*. Ottawa: Canadian Hospital Association Press.

Deber, R. and Leatt, P. (1987). Technology acquisition in Ontario hospitals: you can lead a hospital to policy, but can you make it stick? In: J.M. Horne (Ed.), *Proceeding of the Third Canadian Conference on Health Economics*; Winnipeg: University of Manitoba Department of Social and Preventive Medicine, pp. 259 –77.

Detsky, A.S., O'Rourke, K., Naylor, C.D., Stacey, S.R. and Kitchens, J.M. (1990). Containing Ontario's hospital costs under universal insurance in the 1980s: what was the record? *Canadian Medical Association Journal*, **142**, 565–72.

Evans, R.G. (1992). Canada: the real issues. *Journal of Health Policy, Politics and Law*, **17**, 4, 739–62.

Evans, R.G., Barer, M.L., Stoddart, G.L. and Bhatia, V. (1994). Who are the Zombie masters, and what do they want. Paper presented at Toronto Premier's Council on Health, Well-being and Social Justice. Toronto: Government of Ontario.

Eyles, J. and Birch, S. (1993). A population needs-based approach to health-care resource allocation and planning in Ontario: a link between policy goals and practice. *Canadian Medical Association Journal*, **84**, 112–17.

Feeny, D. (1994). Technology assessment and health policy in Canada. In: A. Blomqvist and D.M. Brown (Eds), *Limits to Care: Reforming Canada's Health System in a Age of Restraint.* Toronto: C.D. Howe Institute Policy Study 20.

The Globe and Mail (1995). Cash will cut waits for surgery. 2 March.

Government of British Columbia (1991). *The Report of the British Columbia Royal Commission on Health Care and Costs: Closer to Home.* Victoria: Government of British Columbia.

Government of British Columbia (1993). *Healthy Communities*. Victoria: Government of British Columbia).

Government of Canada (1969). *Task Force Reports in the Cost of Health Services In Canada.* Edmonton: Queen's Printer.

Government of Canada (1984). *Canada Health Act.* Ottawa.

Government of Canada (1992a). *Strategic Directions for Canadian Physician Resources Management*. Ottawa: Provincial/ Territorial Conference of Ministers of Health.

Government of Canada (1992b). *Federal-Provincial Study on the Cost of Government and*

Expenditure Management. Ottawa: Department of Finance.
Government of Canada (1994a). *When Less is Better: Using Canada's Hospital Efficiently*. Ottawa: Conference of Federal / Provincial / Territorial Deputy Ministers of Health by the Working Group On Health Services Utilization.
Government of Canada (1994b). *An Examination of the Incentives and Discentive Inherent in the Funding Mechanisms of the Canadian Health Care System*. Ottawa: Health Canada.
Government of Canada (1995). *Budget In Brief*. Ottawa: Department of Finance.
Government of Manitoba (1993). *Quality Health for Manitobans – The Action Plan, Partners for Health*. Winnipeg: Manitoba Health.
Government of New Brunswick (1992). *An Act to Amend the Hospitals Act*.
Government of Saskatchewan (1993). *An Act Respecting Health Districts*.
Hastings, J.E. and Vayda, E. (1986). Health services organization and delivery: promise and reality. In: R.G. Evans and G.L. Stoddart (Eds), *Medicare at Maturity: Achievements, Lessons and Challenges*. The Banff Centre for Continuing Education, pp. 337–84.
Hurley, J., Lomas, J. and Vandna, B. (1994). When tinkering is not enough: provincial reform during the early 1990s . *Canadian Public Administration*, **37**, 3.
Jacobs, P., Bay, K. and Hall, E.M. (1993). Inpatient case mix funding in Alberta: an overview. *Proceedings of the Fifth Canadian Conference on Health Economics*. Regina.
Kahn, K., Rubenstein, L.V., Draper, D., Kosecoff, J., Rogers, W.H., Keeler, E.B. and Brook, R.H. (1990). The effects of the DRG-based prospective payment system on quality of care for hospitalized Medicare patients: an introduction to the series. *Journal of the American Medical Association*, **264**, 1953–5.
Linton, A.L. and Peachey, D.K. (1990). Guidelines for medical practice: the reasons why. *Canadian Medical Association Journal*, **143**, 485–90 (part 1), 629–32 (part 2).
Lomas, J. and Barer, M.L. (1986). The legacy of Canadian health care policy. In: R.G. Evans and G.L. Stoddart (Eds), *Medicare at Maturity: Achievements Lessons and Challenges*. The Banff Centre for Continuing Education.
Marleau, Diane (1995). Minister of Health, Government of Canada, as quoted in a letter to the Editor by the Paul Ramsey Minister of Health Province in British Columbia. *The Globe and Mail*, 13 February.
Mickleburgh, R. (1993). Study calls ECG waste of money. Toronto: *The Globe and Mail*, 12 May, p. A1.
Nestman, L.J. (1970). An examination of the economic and administrative structure of health care plans in Canada. Thesis. Edmonton: University of Alberta..
Nestman, L.J. (1989). *Management Control and Funding Systems: For Canadian Health Service Executives*. Ottawa: Canadian College of Health Services Executives.
Nestman, L.J. (1990). Nova Scotia's health system in a crisis. *Proceedings of the Cuban Conference on Health Care Systems in Economic Crisis*. Washington DC: Pan American Health Organization.
Nestman, L.J. (1995a). The Canadian and American health care systems: different values, different funding systems, different health care policies. *Proceeding of the Third and Fourth Meeting of the Expert Network on Health and Health Care Financing Strategies. World Health Organization Regional Office for Europe*. Antidotum: Management in Health Sector, Supplement No. 1.
Nestman, L.J. (1995b). New roles: the Canadian health care system in flux. *Transition*. The Vanier Institute of the Family, Ottawa, June.
Plain, R.H. (1993). Teaching old dogs new tricks: the problems encountered in attempting to reform the method of paying Alberta Hospital. *Proceedings of the Fifth Canadian Conference on Health Economics*. Regina.
Rachis, M. and Kusher, C. (1994). *Strong Medicine: How to Save Canada' Health Care System*. Toronto: HarperCollins.
Redelmeier, D.A. and Fuchs, V.R. (1993). Hospital expenditures in the United States and Canada. *New England Journal of Medicine*, **328**, 772–8.

Schurman, D.P. and Lynch, T. (1994). Reorganizing an academic medical center in a period of change. *Quality Management in Health Care*, **3**, 1.

Soderstrom, L. (1994). Health care reform in Canada: restructuring the supply side. In: A. Blomqvist and D.M. Brown (Eds), *Limits to Care: Reforming Canada's Health System in an Age of Restraint*. Toronto: C.D. Howe Institute Policy Study 20.

Stoddart, G.L., Barer, M.L. and Evans, R.G. (1994). User charges, snares and delusions: another look at the literature. Paper presented at Premier's Council on Health, Well-being and Social Justice. Toronto: Government of Ontario.

Taylor, M.G. (1986). Insuring national health care. In: R.G. Evans and G.L. Stoddart (Eds), *Medicare at Maturity: Achievements, Lessons and Challenges*. The Banff Centre for Continuing Education.

Taylor, M.G. (1987). *Health Insurance and Canadian Public Policy: The Seven Decisions That Created the Canadian Health Insurance System and Their Outcomes* (2nd edn). Montreal: McGill-Queen's University Press.

Tholl, W.G. (1994). Health care spending in Canada: skating faster on thinner ice. In: A. Blomqvist and D.M. Brown (Eds), *Limits to Care: Reforming Canada's Health System in an Age of Restraint*. Toronto: C.D. Howe Institute Policy Study 20.

Chapter 10

Swedish Experience with Fixed Regional Budgets

Johan Calltorp

INTRODUCTION

In the predominantly public Swedish health care system a successful cost control policy was executed during the 1980s. This was made possible by a long tradition of budget planning and steering at all levels of the system – clinical departments and outpatient settings, hospital and county council level. An important dimension of the system is regionalized planning of highly specialized care. It is not unique in Europe but it has been successively developed and deepened since the early 1960s. These factors form a base for the quite remarkable constraint of health care spending that has been achieved during the later part of the 1980s and early 1990s. Parallel to this, tensions have increased throughout the entire health care system because of the growing demands for health care.

During the first half of the 1990s, attempts were made in a number of county councils to introduce a purchaser–provider split and internal market mechanisms. Productivity has increased because of – or associated with – these new steering mechanisms. Several new policy issues also have emerged such as prioritizing needs. Conflicts, tensions and ethical dilemmas become more visible on the clinical level where the cost constraints are transformed to clinical decisions regarding individual patients. Demand for health care will grow considerably, as in all other western countries, mainly due to technological development and demographic transition. A future policy problem will be to maintain stable cost control as market mechanisms are introduced into the system.

GENERAL ASPECTS OF THE SYSTEM

The Swedish health care system is predominantly (90%) public in all its aspects – financing, ownership of health care facilities and control of the system. Sweden has, among the Nordic countries, developed the most comprehensive public structure,

Fixing Health Budgets: Experience from Europe and North America.
Edited by F. W. Schwartz, H. Glennerster and R. B. Saltman. Technical editor R. Busse.

while countries such as Norway and Denmark have systems with the same degree of public financing, but with a somewhat greater proportion of private providers of care (Rohde et al., 1986).

The Swedish system can also be characterized by a balance of power and responsibilities between local and central actors in the system. The 26 local county councils are responsible for provision of the whole spectrum of health services (hospital care, outpatient services as well as primary care). Local political control is exercised by politicians elected every third year with an overall responsibility for running the county councils. An important part of the local power base is the county councils' right to levy taxes. Approximately 80% of total health care funding is collected this way. Patients' direct payments still form a minor part, although direct user fees have been increased considerably in recent years. There is an annual maximum direct payment of 1800 SEK (about $250) per patient. The remaining component is a national government contribution to county councils given as block contributions according to a population-based formula with adjustments for socioeconomic factors similar to the British RAWP (Resource Allocation Working Party) formula. As described later, this has been one of the government's most forceful tools in exercising its successful cost containment strategy. Other steering mechanisms used by the national or county councils have been traditional policy programs, manpower planning and control (especially regarding physicians) as well as traditional consensus building activities.

Each county council is responsible for health care for the population within a defined geographic area, normally between 200 000 and 400 000 inhabitants. A typical county council contains two to four smaller general hospitals and one main hospital with the majority of medical specialities. In addition the country has since the 1960s been divided into medical regions (currently six regions) with one or two tertiary care hospitals. The initial regional structure was developed further with detailed planning and cooperation regarding the particular medical conditions taken care of and treatment protocols at various levels of the delivery system.

PRIMARY CARE – THE MISSING LINK?

Since the 1950s the Swedish health care system has been heavily hospital oriented with a dominance of specialized inpatient services. Earlier, during the first part of the twentieth century, outpatient services given by *provinsialläkare* (district physicians) with a strong public health perspective formed a dominant part of the system. The reasons why specialized hospital services became so dominant from the 1960s onwards has been discussed by several authors (Borgenhammar, 1968; Heidenheimer, 1980). A special climate and value structure seems to have been present during the early part of the 1960s when the postwar system took its major form. There was a strong belief in biomedical advances linked to "social engineering" concepts in the welfare state. Politicians were interested in building hospitals, which were more visible than primary care activities. During a long period Sweden differed from all neighbouring Nordic countries and most other countries in western Europe in the lack of a comprehensive well functioning primary care system (or gate-keeper function). Numerous efforts have been made since the early 1970s to strengthen the primary care component of the system, but with poor results. Primary care has been organized in *vårdcentraler* (primary care centers) with physicians in groups, nurses and auxiliary personnel. There has been constant criticism of the functioning of this system with access problems and too little choice for patients (Saltman and von Otter, 1987). A proposal was put to parliament, and adopted in 1994, to introduce the concept of a general practitioner system with the population listed with GPs and a capitation payment – a system similar to the British. This system began to be implemented but this was reversed when a new government in 1995 decided to make the new arrangements voluntary. At present primary care organization is not uniform and seems to be undergoing change with a considerable variation over the country.

PHYSICIAN AND MANPOWER POLICIES

During the great expansion period in the 1960s and 1970s when health care resources as a percentage of GNP increased from about 4% to a peak level of 9.7% in 1982 (Calltorp, 1989) manpower issues were of great national policy interest in Sweden. The planning and "distribution" of health personnel – especially physicians – was regarded as a key issue in a coordinated and controlled development of the system. Programs were developed with a "planning horizon" ahead in time with a desired distribution of physicians over the medical specialities. Training posts and posts for fully licensed specialists were distributed geographically and over the medical specialities. This system, in effect until the late part of the 1980s, partly succeeded in solving balance issues between the different medical specialities – except for primary care – but did not solve the geographical distribution problems (Calltorp, 1990). This system could also be regarded as one of the national government's most important tools for steering and controlling the county councils. Gradually a constant lack of physicians with many vacancies was turned into a more balanced situation around 1990. The need and use for this type of steering mechanism was no longer at hand and county councils at that time also demanded greater freedom from government regulations as a part of a general decentralization process. The extensive manpower planning system, that had its origin in the early 1970s, was abandoned. At present, Sweden has a relatively high ratio of physicians (1/350 inhabitants) and, from 1994, around 400 physicians were reported unemployed (Swedish Medical Association, 1995).

Physicians have been salaried since 1970 when all private components of the payment system disappeared with the "Seven Crown reform" (Calltorp, 1989). The different medical specialists have been equally paid and wage differentials are generally small. In recent years there has been increasing interest in using wage differences to promote productivity and efficiency among physicians.

PLANNING, BUDGET STEERING AND DECENTRALIZATION

The budget mechanism has a long tradition of serving as a steering and control function in the Swedish system. The system has always been heavily public regarding financing. Technically Sweden has a "health insurance" but that is by law defined so that the County councils have the obligation to provide all needed health services for the population in their respective geographic area. Different phases in the development of today's comprehensive system could be described as the following (Calltorp, 1989):

- 1950s was the phase when formal health care insurance was introduced by law and linked to the county councils that earlier organized only hospital services
- 1960s was the phase of hospital building, structural planning and the beginning of a regionalized system. County councils are by law obliged to cooperate to maintain and finance the regional hospital(s) in their respective areas
- 1970s was a phase of manpower planning and its linkage to health policy program planning, a consolidation of the public system where county councils were placed in charge of running all parts of the system. The government that had earlier provided some health services (psychiatry and primary care) left these to the county councils
- 1980s was the phase when resource expansion came to a halt and mechanisms of constriction and constraint of the system started
- 1990s saw increased tensions in the system as a result of financial constraints and the introduction of some types of market-oriented steering mechanisms, mainly along the lines of the internal market in Britain (a purchaser–provider split). These developments are, however, not uniform over the county councils.

The development over time can thus be described as an evolution in which the county councils have become the deliverers of all health services while the government (which earlier organized primary care and psychiatry) since the 1970s has concentrated on overall planning, controlling and steering activities. The county councils have their power base in the local taxation which collects 80% of the financing. The national government and parliament have the authority over laws and general regulations. Since the mid 1980s the government has generally lost several of its earlier more detailed steering mechanisms, such as the manpower planning with the linked instrument of permission for each new medical post. Several earlier areas of control and steering have been transformed into more general rules and regulations. This is a trend toward decentralization that has evolved in all parts of Swedish society (Bladh, 1987).

Within the county councils the management process too has been evolving in the same direction. Steering and control has consistently been linked to the budget mechanism. On the county council level, a yearly budget decision has been made defining the spending limits for all parts of the system. These budgets have traditionally been quite detailed and have been used as the central mechanism in the system. During the 1980s decentralization was introduced also within the county councils in an effort to increase efficiency and flexibility. Budget mechanisms were developed which included only the broad frames of spending for a specific part of the system, a clinic or primary care center. Generally, personnel budgets have been kept separate allowing more possibility for central influence.

This whole series of developments has encouraged more flexibility and responsiveness in the system to patients (customers) as well as other organizational parts of the entire health care system. This was the first and very important step in a process of changing and "loosening up" a quite rigid and according to some observers "frozen" health care delivery system (Saltman and von Otter, 1987). The combined effects of the public dominance of the system, the budget control and steering and the emphasis on planning had until the mid 1980s created a system that seemed quite rigid. A lot of the scholarly, professional and public discussion focused on these characteristics of the system.

HOW SUCCESSFUL COST CONTAINMENT WAS ACHIEVED

Until the beginning of the 1980s, Sweden allocated a steadily increasing fraction of its GDP to health care services. 1982 marked the peak year with 9.7% of GDP spent on health services (Calltorp, 1989). The pattern is similar to most industrialized countries, although most western countries started to contain costs somewhat earlier (Schieber and Pouillier, 1987). This was in line with general government policy in Sweden at that time to maintain economic growth by increased government spending.

The goal since 1982 has, on the contrary, been to achieve a steady decrease of health spending as a percentage of GDP. At present the reported figure is approximately 7.5% (OECD, 1994). Sweden and Ireland are the only OECD-countries reporting a two digit (over 10%) decrease of health spending as fraction of GDP since the mid 1980s. When the decrease in reported spending from 9.7% to 7.5% is analyzed, it can be seen that around 1–1.5% of the decrease could be accounted for by transferral of non-medically defined long-term care to the municipalities in 1992. It is nevertheless evident that Sweden has succeeded in one of the most pronounced cost containment programs in health care at a system level. How has this been achieved?

Since 1982, when the general economic situation demanded restraints, successive waves of cost containment have been executed throughout the system. The national government has taken a lead role in forcing the county councils to contain costs. One important instrument, for example, has been an agreement between the government and county councils not to raise taxes and a redefinition of the government's contribution to the county councils (Calltorp, 1989). During the 1980s formal annual negotiations around these issues took place between the parties.

The outcome of the negotiations between the Government and the county councils (through the Federation of County Councils) influenced the general budgets through the steering mechanisms described earlier. During the 1980s the reductions were, by these instruments, transformed into general savings campaigns, wage freezes and decreased budgets for investment in equipment and buildings. There seemed to be, at that time, room for increased organizational efficiency without any major interference with clinical activities.

By 1990, pressure had accumulated in the system partly because no major structural reforms had been undertaken during the first cost containment phase in the 1980s. Relative overcapacity of hospital beds in certain areas and imbalances between different segments of the system had emerged. These imbalances made shortage of resources visible in one area and relative oversupply apparent in others. The search for new organizational methods and the need for a more flexible and responsive health care system were important background factors in the quite rapid evolution of different types of purchaser–provider arrangements similar to the UK development of the internal market in the NHS (Ham, 1994). There were, however, other inputs and stimuli for reforms in Sweden. The obvious need for greater flexibility and new mechanisms for resource allocation and reallocation in a strongly cost contained system was one driving factor. Structural problems had accumulated towards the end of the 1980s with imbalance between acute and long-term care. Active observers (Saltman and von Otter, 1987; von Otter and Saltman, 1990) who argued for a more responsive system and the need for introducing greater freedom of choice from the patients' perspective were also of great importance. Other factors of relevance for the reform pattern that developed in the 1990s were the lack of real incentives for efficiency in the production system until that time. This can be described as a consequence of the budget system and the way it had been performed in Sweden without, for example, any stimulus for productivity within the global budget. Calculations of conventional productivity showed a decrease on an average 2% between 1972 and 1982 (Lindgren and Roos, 1985).

DEVELOPMENT OF INTERNAL MARKET MODELS

Starting in 1990 in the county council of Dalarna, a wave of transformation began in the county councils of Sweden from the traditional budget steering model to some form of organizational split between purchaser and provider. An overview has recently been published (Bergman, 1995). About half of the county councils have moved in this direction while the others have maintained traditional methods of steering. Typical is the new diversity – in an earlier very homogenous system – that has now occurred in this sector. The county councils are marketing their changes and reforms under individual names like "the Dala model", "the Stockholm model", "the Bohus model", and so on. These new arrangements all have in common some form of separation between purchasers and providers, although the detailed arrangements differ.

Purchasing could be centrally located in the county council ("Sörmland model") which gives a close resemblance to the budget model, it can be placed distant with comparatively great autonomy and linked to primary care ("Dala model"). In the County council of Stockholm (population 1.8 million) the most far-reaching purchasing–providing model was launched in 1992. Budgets were taken away from all hospitals and were placed in the hands of ten geographically defined population areas that were given the task to develop contracts with the hospitals. Population needs from a public health perspective as well as a disease-oriented perspective were meant to be the basis for the contracting. In addition to the difficulty of rapidly developing the kind of expertise and skills needed for purchasing or "commissioning" health services, the politicians had permitted patients' freedom of choice of hospital to override the contracts. This was meant to put extra pressure on both providers of care and the purchasers – however, one clearly visible effect after two full years of the scheme is the severe problem of handling the oversupply of hospital resources and developing ways to make structural changes to the supply of health care facilities. The immediate effect, even so, has been a marked productivity improvement. Waiting lists that were a constant phenomenon during the 1980s almost disappeared during the first years of the new system (Dahlström and Ramström, 1994).

Another interesting aspect of, especially, the "Stockholm model" with a DRG-based tariff for hospital payment according to the contracts (a technique used by most of the county councils with purchasing–providing) is the rise of new issues in the field of ethics and economics. The incentive structure is markedly changed when clinical departments are remunerated according to each production unit (defined as DRGs) rather than according to a fixed budget. It has been shown that physicians are affected by these "price tags" and perform fewer laboratory tests and diagnostic procedures than before (Calltorp, 1995). The awareness of costs in the clinical meeting between doctor and patient has been increased and with that the discussion has turned to ethical issues and conflicts. There has not, however, been any evidence of medical negligence within the system, and service quality has not been shown to be markedly changed (Forsberg and Calltorp, 1993).

THE WAY AHEAD

The general discussion on ethics, economics and rationing in the health care system of Sweden has been stimulated by big scale changes such as the purchaser–provider split in Stockholm county council. The dynamic that has followed from the marked constraint of health care resources (or spending) since the mid 1980s and the increasing demands triggered by biomedical technology, demography and a general awareness in the population about patients' rights is creating new dilemmas. They are by no means unique for Sweden and are now discussed in many western countries under headings such as "rationing" or "priority setting". In Sweden a parliamentary committee that had been working since 1992 put forward its final report in the spring of 1995 (Swedish Parliamentary Priorities Commission, 1995). This is now widely circulated and discussed in Sweden.

So far the Swedish health care system has succeeded in balancing cost containment with an increased flexibility, increased patient freedom of choices and somewhat increased productivity. The fixed budget mechanism has been an important tool in this achievement and it has partly been redefined regarding its specific function. The way ahead seems to include an even more subtle balancing act between cost control – where the fixed budget seems to be an invaluable instrument – and the flexibility needed for an efficient and consumer-oriented system. The Swedish case illustrates so far that this balance between partly conflicting goals is a difficult act. The next few years will presumably illustrate this even further.

REFERENCES

Bergman, S.E. (1995). *Att beställa hälso- och sjukvård. Erfarenheter från landsting med beställarstyrning*. Landstingsförbundet, Stockholm.

Bladh, A. (1987). *Decentraliserad förvaltning. Tre ämbetsverk i nya roller*. Lund: Studentlitteratur. Akad. avh.

Borgenhammar, E. (1968). *Makten över sjukhuset*. Stockholm: SNS. Akad. avh.

Calltorp, J. (1989). Priority-setting and the decision-making process in health care. Some postwar characteristics of health policy in Sweden. Doctoral dissertation. Uppsala University: Department of Social Medicine.

Calltorp, J. (1990). Physician manpower politics in Sweden. *Health Policy*, **15**, 105–18.

Calltorp, J. (1995). The effects of market mechanisms in a public health care system: the case of Sweden. In: R. Williams (Ed.), *International Developments in Health Care*. London: Royal College of Physicians.

Dahlström, A.L. and Ramström, D. (1994). *Stockholmsmodellen. Effekter, Problem, Vägval. En sammanfattning av ett utvärderingsprogram*. Stockholm: Spri Rapport 398.

Forsberg, E. and Calltorp, J. (1993). Ekonomiske incitacment förändrar sjukvården. (Economic incentives change health care.) *Läkartidningen*, **90**, 2611–14.

Ham, C. (1994). *Management and Competition in the New NHS*. Oxford: Radcliffe Medical Press.

Heidenheimer, A.J. (1980). Conflict and compromises between professional and bureaucratic interests 1947–72. In: A.J. Heidenheimer and N. Elvander (Eds), *The Shaping of the Swedish Health Care System*. London: Croom Helm.

Lindgren, B. and Roos, P. (1985). *Produktions-, kostnads- och produktivitetsutveckling inom offentligt bedriven hälso- och sjukvård. Rapport till expertgruppen för studier i offentlig ekonomi*. Stockholm: Finansdepartementet. (Ds Fi 1985:3.)

OECD (1994). *Annual Health Statistical Report*. Paris: OECD.

Rohde ,T., Garpenby, P., Häkkinen,U., Mabeck, C.E. and Vohlonen, I. (1986). *Privat och offentlig insats i Nordens helsevesen*. Oslo: Gruppe for helstetjenesteforskning, Statens Institutt for folkehelse. (Rapport 3/1986.)

Saltman, R.B. and von Otter, C. (1987). Revitalizing public health care systems; a proposal for public competition in Sweden. *Health Policy*, **7**, 21–40.

Schieber, G.J. and Poullier, J.P. (1987). Recent trends in international health care spending. *Health Affairs*, **6**, 105–12.

Swedish Medical Association (1995). *Report on Medical Manpower*. Stockholm.

The Swedish Parliamentary Priorities Commission (final report) (1995). *Priorities in Health Care. Ethics, Economy, Implementation*. Stockholm: Department of Health.

von Otter, C. and Saltman, R.B. (1990). *Valfrihet som styrmedel*. Stockholm: Arbetslivscentrum.

Chapter 11

Setting Regional Allocations and National Budgets in the UK

Clive H. Smee

INTRODUCTION

The United Kingdom's National Health Service was launched in 1948 with singularly little attention to the financial implications. "Even the Treasury dog did not bark. The assumption was that the cost of the NHS could be calculated simply by extrapolating pre-war health care expenditures ... [and] that expenditure on health care would tend to be self-liquidating by producing a healthier population" (Klein, 1983). Subsequent experience showed how wrong these assumptions were. Health care expenditure in Britain has been subject to the same inexorable upward pressures as in other countries. In many eyes Britain has been uniquely successful in resisting these pressures. Despite, or perhaps because of, this success, debates about the adequacy of funding of the NHS have been a recurrent feature since 1948. For the latter half of this period there has been nearly as much political interest and perhaps rather more academic involvement in how NHS resources should be allocated across the country.

The purpose of this chapter is to set out the main factors that have entered into arguments over the total size of the NHS budget and its regional allocation; to summarize what we have learned about the importance of these different factors; and to suggest some outstanding issues. The first part looks at the determinants of the NHS budget. The second part is concerned with the methods of allocating that budget to regional authorities. The evidence quoted refers to England only, though similar arguments will usually apply to the other countries of the UK.

DETERMINING THE NHS BUDGET

A Short History

For the first two years of its existence NHS hospital expenditure was essentially determined by the bids of individual hospital authorities. This bottom-up system

Fixing Health Budgets: Experience from Europe and North America.
Edited by F. W. Schwartz, H. Glennerster and R. B. Saltman. Technical editor R. Busse.

quickly generated rapid and unexpected increases in expenditure and from 1950–51 a global budget limit was placed on hospital expenditure, both current and capital. The "exercise of financial control through global budgets", to quote the architect of the NHS, Aneurin Bevan, became a guiding principle of the NHS (Klein, 1983).

But there were and continue to be exceptions. The top-down limit on expenditure set by the Ministry of Health in 1950 related to only two-thirds of the total health budget. The rest, broadly primary care services and drugs, was effectively left open-ended. Parts of these programmes were subsequently brought within budget caps but major elements still remain essentially demand-led and deserve separate discussion.

A second major date in the history of NHS budgets is 1976 when the cash-limit system was introduced. Before that date the top-down budgets set for the NHS as a result of annual discussions between the health department and the Treasury were based on certain assumptions about inflation and were subsequently adjusted for actual changes in price levels. Since 1976, budgets have been set on the basis of Treasury determined assumptions about inflation and there has been no adjustment if price increases have proved to be higher or lower than expected. In some years additional allocations have, however, been made to meet the cost of unforeseen increases in salary and wage rates.

The system of global budgets and its increasingly tight enforcement has undoubtedly contributed to the relatively slow growth in UK health care expenditure since the 1950s. But other aspects of the UK system of funding health services have also been important. Because the NHS has been centrally funded, essentially from national taxes not national insurance contributions, the Treasury has always been able to scrutinize and challenge the allocation of public funds for health service purposes. Over time it has become increasingly proficient in questioning not only the need for additional services but also the cost-effectiveness of existing services. The annual budget-setting negotiations between the Department of Health and the Treasury are now supplemented by a year-round dialogue on the scope for improving value for money. In addition to cascading the centrally agreed global budgets to lower administrative levels the Department of Health has long had a range of tools for influencing the quantity and price of services provided. It plans the supply of health service manpower and can influence wage levels; it essentially determines the size and allocation of capital expenditure and hence the number of hospital beds; it also has a range of mechanisms for influencing drug prices and total expenditure on pharmaceuticals.

These powerful and effective means of controlling health costs have not stopped total health spending from growing rapidly in absolute terms. For example, between 1978–79 and 1994–95 expenditure on the NHS grew twice as fast as both national income and general government expenditure as a whole: see Figures 11.1 and 11.2. But compared with other western countries the growth of total health expenditure and public health expenditure (85% of total expenditure) has been relatively modest. In every decade since 1960 the elasticity of total expenditure on health service growth

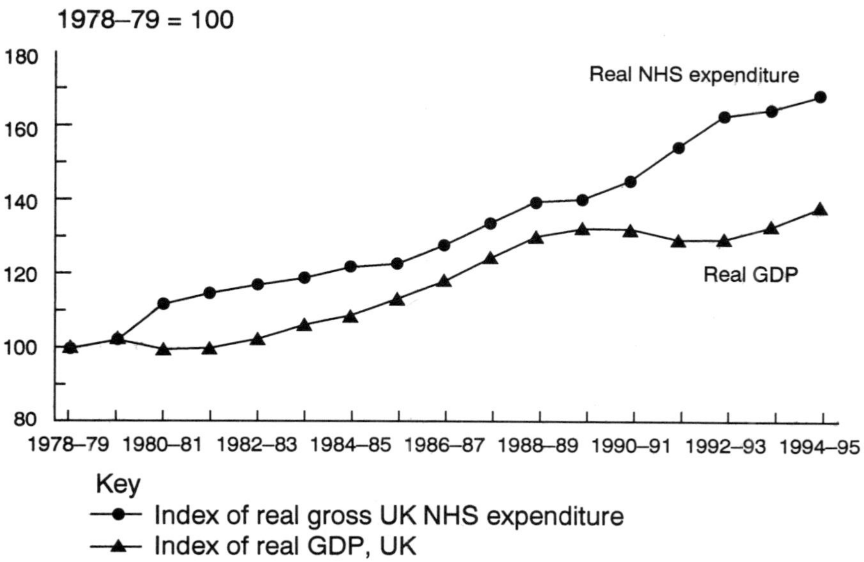

Figure 11.1 Growth in GDP and NHS expenditure, 1978–79 to 1994–95

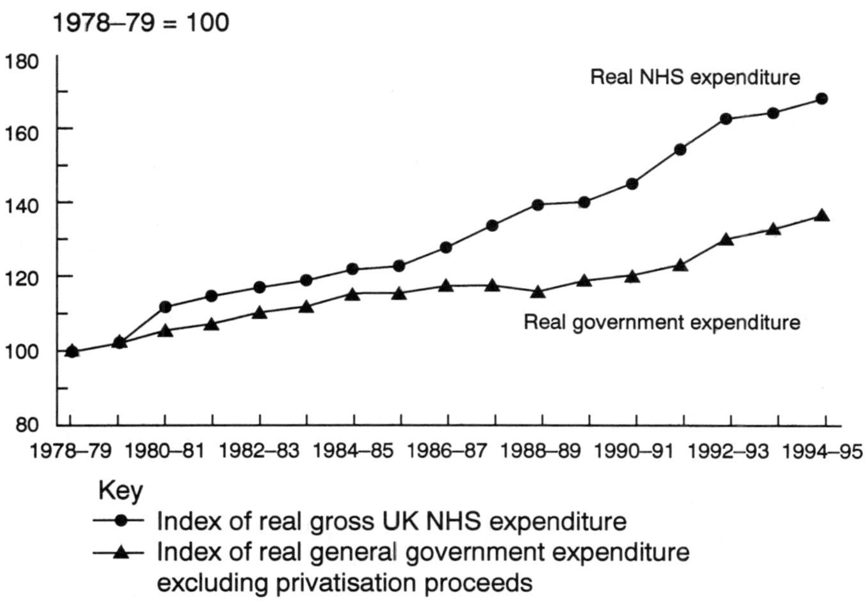

Figure 11.2 Growth of general government expenditure and NHS expenditure, 1978–79 to 1994–95

to total domestic expenditure growth has been roughly 0.1% per annum below the European average (OECD, 1993). As the UK obtains health outcomes that are broadly comparable with those achieved in other countries the NHS has been judged by the OECD to be "a remarkably cost-effective institution" (OECD, 1994, p. 64).

Determinants of Capped Budgets

Hospital Current Expenditure

Since the 1950s all governments in Britain have recognized that how much should be spent on hospital services in particular and on the NHS in general is essentially a political judgement. In 1953 a committee of inquiry into the cost of the National Health Service (Guillebaud Committee) was asked to advise on what would be involved in "providing for the maintenance of an adequate service". The Committee's report concluded that there was no way of setting a long-term financial target which would ensure that an "adequate service" would be provided. The report went on:

> We conclude that in the absence of an objective and obtainable standard of adequacy the aim must be, as in the field of education, to provide the best service possible within the limits of the available resources. It is clear that the amount of national resources expressed in terms of finance, manpower and materials which are to be allocated to the National Health Service must be determined by the Government as a matter of policy, regard being had to the competing claims of other social services and national commitments, and to the total amount of resources available. (The Guillebaud Committee, 1956, para. 98)

No subsequent government committees have been asked to advise on the resources required for an "adequate service" though this has not discouraged policy analysts and politicians from making their own judgements on the adequacy of funding. The 1978–79 Royal Commission on the NHS did ask itself this question. It concluded: "There is no objective or universally acceptable method of establishing what the 'right' level of expenditure on the NHS should be" (Royal Commission, 1979).

The acceptance of all governments since Guillebaud that budgets for the NHS were bound to be policy judgements and could not be left to be determined by objective measures of "adequacy" has not stopped the search for objective factors that should influence those judgements. Factors frequently quoted in debates about pressures on NHS budgets have included: demography; public expectations/maintaining service levels/unmet need; medical advance; new threats to public health; relative price effects; scope for efficiency savings; reform "investments"; and "political considerations". A detailed review of the relative importance attributed to these factors and how they have been (or should be) measured, is beyond the scope of this chapter, but some short comment may be helpful.

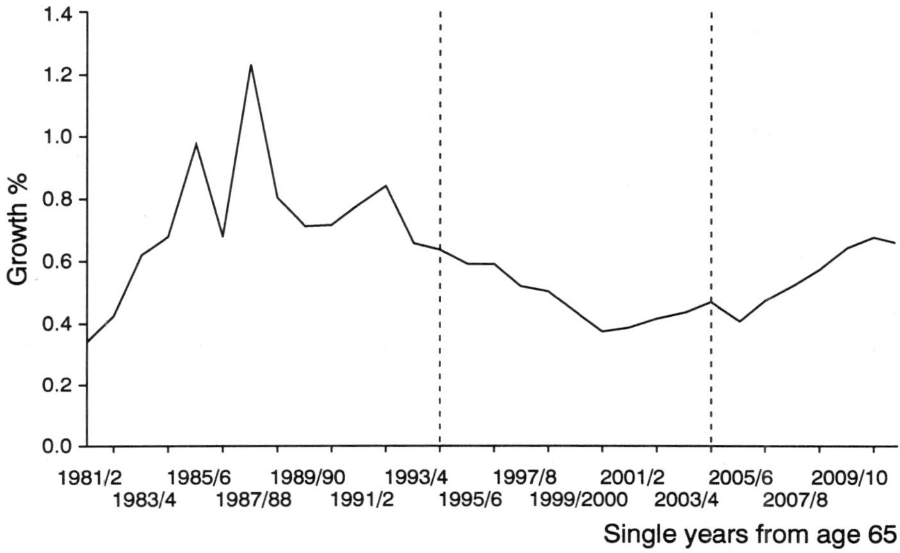

Figure 11.3 Effects of demographic change – HCHS

Demography

This factor has been studied more carefully than any of the other possible pressures on the NHS budget. A study in the 1950s by Abel-Smith and Titmuss (1956) estimated the effect of projected age changes in the population on the cost of the National Health Service over the next 20 years. They concluded that demography alone would add about 8% to the cost of the NHS, or less than 0.5% a year. Current estimates, using much better information than was available to Abel-Smith and Titmuss, suggest that demographic changes are now adding to the pressures on the NHS by between 0.4% and 0.6% per annum – see Figure 11.3. Conventionally, estimates of the impact of demography have assumed that present expenditures across age groups are adequate and the relativities stable. But it is possible to make more refined calculations allowing for changes in utilization rates between age groups.

In 1956 Abel-Smith and Titmuss concluded that other social factors, notably changes in the proportion of men and women who are single, widowed, divorced and childless might be more important factors in influencing future demand for medical care than any foreseeable changes in age structures. There is some evidence that numbers of single parents and elderly people living alone or with single carers are reliable indicators of need for hospital services but there have been no recent projections of their possible impact.

Public Expectations

All governments recognize that the populations' demand for health care rises through time with increases in income and wealth, i.e. there appears to be a positive income elasticity. The UK government of the early 1990s, for example, has a manifesto commitment to "year by year, increase the level of real resources committed to the NHS" (Conservative Central Office, 1992). Various ways have been suggested of trying to monitor and quantify public expectations. One approach is to look at the recent growth in health service volumes sector by sector and assume that public expectations require recent trends to be at least maintained. For some years the Department of Health has used a simple model that builds on this assumption (see Appendix 1). A weakness of this approach of course is that trends in services reflect the interaction of demand and supply, not the shift in demand alone. To try to make some allowance for this, an analysis of changes in service levels can be supplemented by an analysis of trends in hospital waiting lists and waiting times, seen as rough proxies for unmet demands or expectations. In theory it is then relatively simple to calculate what increases in hospital activity levels are necessary to produce given reductions in long waiting times. In practice, limited understanding of the causes, role and dynamics of hospital waiting lists means that such calculations are fraught with uncertainty. Nevertheless, so long as the public sees hospital waiting times as a major, and perhaps *the* major cause of dissatisfaction with the NHS, health departments can be expected to have regard for the service levels that are projected, together with appropriate management action, to reduce those times or at least prevent them from rising.

Medical Advance

New technologies and new drugs can reduce unit costs as well as increase them. But there now seems to be broad acceptance among informed analysts in most countries that the net effect of medical advance is to increase pressures on health service expenditure through constantly expanding the range of conditions that are treatable. However, it has been difficult to find an objective way to calculate the impact of such developments on the NHS budget. For a period in the 1980s, it was argued that medical advance added 0.5% a year to the pressures on the NHS. But this figure was dropped for lack of supporting evidence.

There is now interest in the approaches used by two US federal agencies, the Prospective Payments Advisory Commission and the Physicians Payments Advisory Commission, in calculating the effect of medical advance on Medicare expenditures. These agencies take a bottom-up approach to estimating the impact of medical advance and look for new technologies that are expected to diffuse rapidly over the next year or two, having already reached at least 5% of the target population (PPAC, 1992; PPRC, 1993). A potential weakness of this approach is the circularity built into the assumption about diffusion rates: expectations about diffusion will be linked to expectations about availability of resources. On the other hand, the approach should help to distinguish rapidly diffusing technologies that are cost-effective from those for which there is little current evidence of cost-effectiveness. To the extent that the number and expenditure implications of the former category can be shown to change through time, the bottom-up approach should provide more robust evidence on the need for changes in resources to meet the "pressures" of medical advance.

Changes in Public Health

Evidence of changes in public health would seem the most obvious reason for changes in the NHS budget. Due to a variety of factors, of which improved health services are only one, the general health status of the British population has improved enormously since the NHS was introduced. But this is not seen by health departments as a reason for arguing for fewer resources for health services. The explanation, of course, is simple: the improvement in health status has been accompanied by an increase, not a reduction, in the demand for health services. Under some optimistic (or futuristic) scenarios – e.g. a compression of morbidity – growth in life expectancy could be accompanied by a decline in the use of services among the elderly (Fries, 1980). In England a recent review of the evidence found no support for this hypothesis (Bone et al., 1995). On the other hand new epidemics, such as AIDS, have been accepted as legitimate pressures on the NHS.

Relative Price Effects

A common argument for year on year increases in real resources for health services is that health sector prices rise faster than general inflation, reflecting the labour intensity of the sector. There have been times in the history of the NHS when a great deal of importance has been attached to tracking the price index for (non-labour) hospital services and comparing it with the general price index. While the two can differ significantly in the short term they tend, in the British experience, to follow a similar path over the longer term. With the return to lower inflation this debate has reduced in significance. Separate arguments have been made and continue to be made about the likely trends in health service pay. The willingness of governments to see this grow in real terms has been a function of the stages reached in both the economic and the political cycle. In recent years there has also been an expectation that real improvements in wages should be "earned" through improvements in productivity. In consequence the labour intensity of the health sector is not now seen as an adequate justification for real terms increases in output prices.

Efficiency Savings

While it has probably always been realized that there is scope for improvements in efficiency in the NHS it was only in the mid 1980s that formal allowance began to be made for an efficiency factor in the annual discussions over the health service budget. The discussions were facilitated by the development of a summary measure of trends in efficiency across most of the hospital sector. With the aid of a cost-weighted activity indicator it became possible to track improvements in technical efficiency, always assuming there were no sharp changes in the quality of the activities that were counted. The record revealed was impressive: an average 1.5% per annum increase in efficiency across the 1980s. From the late 1980s public expenditure negotiations assumed that a 1% improvement in efficiency was a reasonable future target. In more recent years this target has been steadily increased. For 1995–96 health authorities have been set a target of an overall 3% increase in efficiency across the hospital and community health services.

The sustained improvement in efficiency over the last 10–15 years raises two questions. First, what have been its sources? Second, can it be continued? The main sources of the reductions in unit costs have been shorter hospital lengths of stay (not just for acute patients but also for mentally handicapped, elderly and mentally ill patients) and the rapid growth of day surgery. Looking to the future, it is not clear that either of these causal factors has yet run its full course.

Investment in Reform

A recent pressure on the NHS budget has been the costs of developing the infrastructure necessary for implementing the internal market announced in the 1989 White Paper on the reform of the UK health service. For three years 1989–90, 1990–91 and 1991–92 the Department of Health agreed with the Treasury that additional resources should be made available for the information technology, training and other start-up costs associated with introduction of the new reforms. This was justified as an "up-front" investment. Unsurprisingly, in more recent years the investment has been expected to yield a return in the form of higher efficiency savings.

Political Considerations

While the above relatively "objective" factors can be used in argument for larger or smaller health service budgets, the annual decision on the NHS budget inevitably represents a political judgement. Some commentators claim that the weight given to "political considerations" (roughly defined as all factors other than those already listed above) depends on the stage of the electoral cycle. But this is far too simple: ministers will be influenced by many considerations. For example, the current government has a manifesto commitment to increase "year on year" the real resources devoted to the NHS. On the other hand, it is also committed to reducing the share of overall public expenditure in national income. Again, governments are well aware that their public expenditure priorities are signalled by the relative rates of growth they accord to expenditures on different programmes. A government that claims to give high priority to health services will therefore want to look at the rate of growth planned for health compared with the growth planned for other areas of public expenditure.

Reflections

Many of the factors seen as important determinants of the NHS budget today were seen as equally important 20 or even 40 years ago. But there have been a number of significant changes, in particular:

- The assessment of each potential pressure has become increasingly sophisticated and quantitative. For example, until recently the effects of demography were calculated using the number of occupied bed days by age band, but now the occupied bed days are split between hotel and treatment costs. This captures the fact that younger people occupy beds for a shorter length of stay but receive more intensive treatment.
- Improvements in efficiency are now recognized as an important downward pressure on budgetary requirements. Indeed, this factor has become a major influence on budget settlements.
- The implicit assumption that baseline budgets were well spent in terms of the effectiveness of care has come under increasing challenge. In future those arguing for increased health service resources in Britain will be expected to demonstrate not only the cost-effectiveness of proposed *new* interventions but also that *existing* resources are going to treatments that are effective and appropriate. Given the difficulties of demonstrating the latter, there will be pressures on health authorities to show that steps are being taken to identify and reduce ineffective treatments and to concentrate resources on effective care.

Hospital Capital Expenditure

Hospital capital expenditure has ranged from about 4% of hospital current expenditure in the early years of the NHS to over 12% in 1972/73. At present, in the mid 1990s, the figure is close to 10% of current expenditure. Interestingly, one study concluded that in the late 1930s, before the founding of the National Health Service, capital expenditure represented nearly 20% of hospital current expenditure (Abel-Smith and Titmuss, 1956).

Both in bidding for funds and in making allocations, capital budgets in the NHS have always been kept separate from budgets for current expenditure. But it has not been possible to develop a wholly satisfactory way of objectively determining the correct balance between the two budgets. On the one hand, there have been reports since the founding of the NHS of buildings of sub-standard quality and of hospital building plans that have been delayed for lack of funds. On the other hand, until 1991 hospitals paid no capital charges and there were common stories of under-used buildings. The introduction of capital charges (to reflect the real cost of capital) and of financial targets and self-governing (Trust) status for hospitals should provide a better guide than ever before to what is the efficient balance between capital and current expenditure. Initial signs are that the introduction of capital charges and of internal markets may have reduced the demand for capital: for example, some major new capital schemes have been reappraised and reduced in size. But given that demand has historically run well ahead of available funds, it is still too early to say whether in the long term there will be any significant downward shift in the ratio of capital expenditure to current hospital expenditure. However, the relaxation of the former very strict rules on access to private capital almost certainly implies an increasing role for private capital.

The NHS is also allowed to reinvest the proceeds from its sales of surplus land and buildings. The value of these sales has been considerable in recent years (for example, as high as £200 million per annum) though annual revenues have been sensitive to changes in the commercial property market.

Determinants of "Demand-led" Budgets

One-sixth of NHS expenditure is still described as "non-cash limited", meaning that it is not subject to capped budgets defined in cash terms. The expenditures in this category are the bulk of the resources going to primary care services (called Family Health Services). The three biggest elements are pharmaceuticals prescribed by primary care doctors (GPs) and the costs of primary care doctors and of primary care dentists. Expenditure on these services remains formally uncapped because there are believed to be strong health arguments for allowing open access to primary care professionals, and because ministers have given a commitment that patients will receive the drugs they need. Given public expectations, any explicit rationing of services in these areas could be unacceptable. The relative success of other types of controls has been a further contributing factor.

In the absence of simple global budgets, governments have attempted to influence and control expenditure on the Family Health Services in other ways. Pharmaceutical cost-containment strategies have included:

- a voluntary scheme of profit-rate controls (the Pharmaceutical Price Regulation Scheme)
- co-payments – currently for less than 20% of prescriptions, and levied as a flat rate charge
- a negative list of drugs not covered by the NHS
- GP education and feedback of prescribing behaviour relative to peers
- encouragement of generics
- indicative budgets for GPs and incentive schemes.

These policies provide governments with a range of potential levers to pull when expenditure on pharmaceuticals threatens to increase at a rate judged to be excessive. Together these policies have delivered prescribing behaviours that are conservative and drug expenditure levels that are low relative to most other countries. Nevertheless, over recent years drug expenditure has been rising at over 4% per annum in real terms, faster than any other element in the NHS budget.

The introduction of "fundholding" for GPs (the granting of budgets to general practitioners) offers a further tool for controlling the growth of pharmaceutical expenditure. The budgets given to "fundholders" include the drugs prescribed by GPs. The budgets are set historically in the first instance but are subsequently cash limited. As fundholding expands, expenditure on prescribed drugs is thus gradually brought within cash limits. This is acceptable because doctors can if necessary make up any shortfall on their drugs budget by viring from other kinds of health expenditure. Currently about half of all GPs are fundholders.

Expenditure on the two other main elements of the non-cash limited services, doctors and dentists, has also risen rapidly in recent years. Control is exercised through setting income targets for practitioners (doctors and dentists) and through influencing the supply of practitioners. The income target for, say, general practitioners is based on the recommendations of an independent pay review body. General practitioners are paid through a mixed system of capitation payments, allowances and fees for service. The "prices" for these various elements are set at a level that is expected to deliver the agreed target income for the average general practitioner. If miscalculations or behavioural changes cause the target income to be overshot, the overshoot can be clawed back in subsequent years. Provided the elements in the mixed remuneration system are well designed, the overall effect is to provide each GP with incentives to improve his or her own performance while maintaining control of total expenditure. The system has worked better for doctors than for dentists. For dentists, consideration is currently being given to how the system should be improved, drawing on an extensive public consultation exercise (NHS, 1994a).

Apart from unanticipated behavioural changes the main way through which expenditures may turn out to be larger than expected is by a rapid increase in the number of practitioners. But the government has a range of levers to influence the growth in numbers. The incentives built into the fee system affect how far it is cost-effective for existing partnerships to take on new partners. And through the relativities between the remuneration of GPs and hospital doctors the government can influence the career choices of those coming out of medical schools. Finally, the total number of doctors in training is also to large measure within government control.

Although the government does not cap expenditure on family doctors, since 1991 there have been formal cash limits on some important aspects of their expenses – premises costs, paramedical staff, and computers. Expenditure on these items now amounts to about £700 million a year in England, or nearly £30 000 per doctor. These sums, which are at the discretion of local health authorities, can be used to help influence the pattern of development of local primary health care services.

ALLOCATING THE NHS BUDGET TO REGIONAL AUTHORITIES

Allocating Current Revenues for Hospital Services

Prior to the establishment of the NHS, health care was financed mainly by private expenditure, charity and municipal finances. In consequence the geographical distribution of expenditure largely reflected disposable income and local taxable capacity. The NHS introduced the principle of allocating expenditure according to the different principle of need. Nevertheless, in the early years of the NHS, revenue funds for hospitals were allocated to regions largely in proportion to expenditure in the previous year. The weakness of this arrangement was recognized early on and the Guillebaud Committee of the 1950s considered "very carefully whether any formula might be devised (related to regional populations, number of beds, and/or other factors) which might serve as a guide to the Health Departments when sharing out the annual allocations to regional boards". The Committee concluded "that any national formula would have to be weighted to take account of such a wide range of variables in hospital regions that it cannot be considered as a practical proposition at least for the present" (Guillebaud Committee, 1956, para. 282). Over the following years there was little progress towards a more equitable basis for regional allocations until an explicit formula (the Crossman formula) was adopted in 1971–72. It was not until 1977–78 that a formula was implemented which set targets for regional allocations mainly on the basis of population need, using weighted capitation principles.

The 1976 report of the Resource Allocation Working Party (RAWP) recommended that resources be allocated to health authorities so as to ensure, over time, equal opportunity of access to health care for people in equal need (DHSS, 1976). RAWP proposed the use of an allocation formula that took account of population size, the use made of health services by people of different ages and sexes, and geographical variations in death rates (as a proxy for variations in morbidity). Adjustments were also to be made for such factors as patients resident in one authority who were treated in another, higher staff costs in London and teaching and research costs. The RAWP proposals were implemented and a target or "fair share" of available resources was calculated for each region using the capitation formula. Allocations to regions moved authorities gradually towards their targets at a pace decided by ministers.

In the decade following the introduction of the formula, disparities in resource allocation between different regions were substantially reduced and RAWP was generally judged to have been a success (Mays and Bevan, 1987). Between 1979 and 1989 the maximum distance above target was reduced from around 13% to around 8%, while the maximum distance below target was reduced from about 9% to 4%. However, as regions moved closer to their targets ministers called for a review of RAWP to explore the scope for improving the way in which the formula measured a Region's relative need for health services. The subsequent report (DHSS, 1988) fed into a revised weighted capitation formula. Regions were recommended to adopt a similar approach in setting hospital and community health service allocations for district health authorities but they were given discretion to add measures of social deprivation.

A further review of the formula was announced in 1993 to take account of new data (from the 1991 Population Census) and of new research work on the determinants of health need. A major analytical study by the University of York led to a number of recommendations for improving the sensitivity to health need of the current formula (Carr-Hill et al., 1994). The main recommendations were accepted at the end of 1994. The principal changes are a more sensitive age/cost curve; the substitution of more direct indicators of variations in morbidity for the standardized mortality ratios that have previously been used as a proxy; and the development of a more embracing "market forces" factor to adjust for variations in factor costs across regions. A description of the modifications and a comparison with the existing formula is in Appendix 2 (NHS, 1994b).

From 1996, subject to Parliamentary approval of the necessary legislation, regional health authorities are to be abolished and the new formula will be directly applied to unified health authorities, which will combine the role of former district authorities and the responsibilities of family health service authorities. (These authorities will on average cover about 500 000 people.)

Reflections

The estimation of needs weights for the population has been increasingly based on empirical work using multivariate analysis of hospital utilization (inpatient admissions) data. Such empirical work has replaced a certain amount of judgement about weights in the early weighted capitation formulae. The difficulty in analysing utilization data is to distinguish variations in utilization due to need from variations in utilization due to supply factors. The trick has been to assemble utilization, supply and need data for "small areas" – local authority wards containing on average about 7500 people each. Since several wards are likely to share access to each hospital it is believed that at such a level the relative utilization between wards will mainly reflect relative need.

Separate analysis is carried out for acute and psychiatric admissions. The needs data includes mortality and a large number of variables from the decennial population census including age, sex, household composition, occupation and housing tenure. In addition, a limiting longstanding illness question was deliberately added to the 1991 census anticipating its use in exercises such as the 1993 review. The supply variables include: hospital beds, distance of the ward from each hospital and access to GPs. The effect of both needs variables and supply variables on need is then investigated using two-stage least squares regression analysis. The needs weights are based on those needs effects which are statistically significant. The supply effects are ignored. The needs variables which emerged as statistically significant in the 1993 analysis are included in Appendix 2.

Distributing Current Revenues for Primary Care

Until very recently the regional distribution of resources for primary care has never been subject to a formula approach. The two recent exceptions are:

- the cash-limit expenditure on family doctor expenses, which is now allocated by a formula similar to that for hospital resources
- the indicative prescribing allocations for GPs, which in 1995, for the first time, took account of "objective" measures of the need to spend as well as historical spending patterns.

Apart from these exceptions resources have essentially followed the distribution of primary care practitioners. A degree of equity has been ensured through the use of a committee, the Medical Practices Committee (MPC), to control the distribution of doctors throughout the country. In assessing the needs of different areas the MPC gives consideration to a range of factors including the average practice list size in the local area, the populations' age composition and geographical spread, special local needs (e.g. ethnic groups, access to women doctors), specific characteristics (e.g. a particularly socially deprived area or a rural isolated community) and the preference of local residents as voiced by Community Health Councils. In recent years the MPC's direct controls have been supplemented by the provision of financial incentives to encourage working in areas of deprivation.

The regional distribution of dentists has generally been left to market forces. By the late 1980s this had led to variations of nearly 1 to 2 in the numbers of people per dentist between major regions of the country. In consequence in 1989 arrangements were introduced for offering financial assistance in the setting up of dental practices in designated areas. However, the scheme has not met its objectives and it is to be scrapped.

From 1996 it is planned, subject to Parliamentary approval, to merge district health authorities and family health service authorities (the authorities responsible for primary care services). This merger will highlight the need to develop a coherent and co-ordinated approach to allocating resources to integrated health authorities. Work is currently being undertaken to address this issue.

Allocating Capital

In the early years of the NHS, capital budgets were allocated to regions mainly on the basis of population but some funds were allocated to hospitals with special needs and a central reserve was maintained for financing large schemes. In 1976 the RAWP report already referred to recommended a new formula for the allocation of capital. Between 1979 and 1984, 85% of capital sums were distributed on the basis of weighted populations and the remaining 15% were used to reduce inherited disparities in the value of regions' capital stocks. Following a review in 1983, when it became clear that the stock equalization factor was not working as intended, it was decided to move towards a system based solely on regions' weighted populations. The current method of allocating capital to regions is to use the same weighted capitation formula as is applied to revenue (excluding "market forces" weighting) – i.e. it takes account only of population, age distribution and need.

OUTSTANDING ISSUES

History would suggest that the size and rate of growth of the NHS budget will continue to reflect political judgements made annually as part of the government-wide debate on public expenditure. There appears to be no interest in replacing the political judgements that come out of the annual negotiation process with some simple rule such as that the rate of growth of expenditure be tied to the rate of growth in national income. But while the process by which the NHS budget is set seems unlikely to change, the factors that are taken into account and their quantification will continue to evolve. Negotiations over future budgets can be expected to continue to grapple with the following issues:

- Can recent rapid improvements in efficiency in the hospital sector be expected to continue? Can they be extended to the primary and community care sectors?
- As more information becomes available on health outcomes and there is greater emphasis on the importance of "evidence-based" health care, will the results of cost-effectiveness analysis support arguments for faster or slower growth in NHS expenditure?
- Will the net effect of medical advance continue to be expenditure-increasing, or will developments such as tele-medicine usher in a new phase in which the net impact is expenditure-reducing? In either case how can the cost impact of new technologies be better measured?
- As interest in the clinical quality of care grows (with better measures of quality) will evening up of quality standards require more resources, or are quality and low unit cost complementary as some US academics argue ? (Fleming, 1991)
- What will be the costs of meeting rising public expectations about the consumer aspects of health service quality, including waiting times and service choice?

Turning to allocating the NHS budget to sub-national units, it seems a safe bet that preference will continue to be given to empirically based formula approaches. But organizational changes mean that this preference faces formidable challenges. In particular:

- The prospective amalgamation of district health authorities and family health service authorities will create pressure to bring the allocation of all health revenues under a coherent formula approach. The problem is that while a formula approach has long been successfully implemented for hospital and community health services no such approach has yet been devised for general practitioners. Indeed, the latter's "demand-led" system of funding is not easily reconcilable with the application of mechanical cash limits.
- The development of GP fundholding poses an even greater challenge to allocation mechanisms. The difficulties with all fundholding budgets are that because they cover small populations and the populations can change through time (including through cream-skimming) there are problems in measuring both systematic and random variations and in distinguishing between patient demand and need. Will it be concluded that pure weighted capitation is not attainable and that there will have to be some supplementary payments according to actual expenditure?
- On the capital side there is a continuing search for allocation mechanisms that will ensure resources are put to best use within the internal market while protecting concerns about geographical equity of access to services.

The challenges faced by the UK's National Health Service in setting overall health budgets and allocating those budgets regionally will be shared, in some form, by many other countries. Hopefully this chapter will contribute to a comparison of experiences and to mutual learning.

NOTE

The views are those of the author only and should not be taken as representing those of his Department or Government.

APPENDIX 1 THE DEPARTMENT OF HEALTH "PES MODEL"

The term "PES model" is used for convenience, but is in itself something of a misnomer since it is not an economic model in the sense for example of the Treasury model. It does not use sophisticated econometric techniques but simply makes a number of separate trend projections of the relevant variables using simple regression (House of Commons Health Committee, 1994). In essence we have a set of linked spreadsheets which:

- take as their starting point HCHS activity and expenditure data for the recent past (data are currently held from 1981) along with corresponding age specific population estimates and projections of population to the end of the survey period. Some of these data have been adjusted to remove discontinuities caused by changes in data collection systems

- use simple regression techniques (with time as the independent variable) to project future activity (or activity rates per head of population) and unit costs on the assumption that future trends are similar to those of the past. The trend can be fitted over a shorter period if required by data discontinuities or if initial trending suggests an underlying change in the trend during the decade or so for which data are held. The model also gives the choice of a logarithmic trend (i.e. one which assumes a constant percentage rate of change) or a linear one (i.e. assuming a constant absolute change)
- multiply up and sum the results of these projections to estimate what overall spending growth would be in future years if past trends continued.

The above description is a slight over simplification for a number of reasons:

- some areas of HCHS activity use age specific activity rates rather than simple rates when the relevant data are available
- for acute and geriatric hospital admissions, unit costs are not projected directly as a function of time. They are assumed to be a linear function (estimated, again, by simple linear regression) of length of stay which is separately projected
- in some areas no consistent relevant activity series is available so a simple projection of expenditure (as a linear function of time) is made.

These projections are just projections of the past trends rather than forecasts of how spending might be expected to change in an unconstrained system or estimates of changes in need or demand for health care. They form the starting point, but only the starting point, for the Secretary of State's judgements and assessments in the public expenditure round.

APPENDIX 2 COMPARISON BETWEEN CURRENT AND MODIFIED WEIGHTED CAPITATION FORMULA

Table 11.1 Elements of the current and modified weighted capitation formula

CURRENT Age/cost curve	MODIFID More sensitive age/cost curve
Need represented by square root SMR< 75	Two need indices: (i) General and Acute (ii) Psychiatric reflecting a number of health and socio-economic variables from 1991 census
Thames Adjustments – London weighting – Thames market forces factor – Thames 3% and 1% supplement	All embracing staff market forces factor (subsumes LW adjustments) based on external comparators recognizing four pay zones
Capital charges separate target calculation using age and need as above plus an MFF reflecting geographical differences in land values and building costs down to RHA level	Totally integrated target calculation using age and need as above, and a modified MFF reflecting geographical differences in land values and building costs down to DHA level

Source: NHS (1994b)

The modified formula weights 64% by the general and acute model, 12% by psychiatric and *gives no weight to 24%* (i.e. applying the York work to 76% rather than 100%). The variables within the two need indices are shown in Table 11.2.

Table 11.2 Variables in the needs indices for general and acute and psychiatric

Need variables	General and acute	Psychiatric
Standardized limiting long standing illness ratio < 75	+	
Standardized mortality ratio (SMR) under 75	+	
Proportion of economically active who are unemployed	+	
Proportion of pensionable age living alone	+*	+*
Proportion of dependants in single carer households	+	
Proportion of persons in lone parent households		+
Standardized mortality ratio (SMR) under 75		+
Proportion of dependants in no carer households		+
Proportion of adult population permanently sick		+
Proportion born in New Commonwealth		+

* These variables are included in both general and acute and psychiatric, but with different coefficients.

REFERENCES

Abel-Smith, B. and Titmuss, R.M. (1956). *The Cost of the National Health Service in England and Wales*. Cambridge: National Institute of Economic and Social Research.

Bone, M.R., Bebbington, A.C., Jagger, C. and Nicholas, G. (1995). *Health Expectancy and its Uses*. London: HMSO.

Carr-Hill, R., Hardman, G. and Martin, S., et al. (1994). *A Formula for Distributing NHS Revenues Based on Small Area Use of Hospital Beds*. University of York. Occasional Papers.

Conservative Central Office (1992). *The Best Future for Britain: The Conservative Manifesto for Britain 1992*. London: Conservative Central Office.

DHSS (Department of Health and Social Security) (1976). *Sharing Resources for England (RAWP Report)*. London: HMSO.

DHSS (Department of Health and Social Security) (1988). *Review of the Resource Allocation Working Party Formula. Final Report by the NHS Management Board*. London: HMSO.

Fleming, S.T. (1991). The relationship between quality and cost: pure and simple? *Inquiry*, **28**, 29–38.

Fries, J.F. (1980). Aging, natural death, and the compression of morbidity. *New England Journal of Medicine*, **303**, 3, 130–5.

The Guillebaud Committee (1956). *Report of the Committee of Inquiry into the Cost of the National Health Service*. London: HMSO. Cmnd. 9663.

House of Commons Health Committee (1994). First Special Report (1993–94 Session): Memorandum from the Department of Health. London.

Klein, R. (1983). *The Politics of the National Health Service*. London: Longman.

Mays, N. and Bevan, G. (1987). *Resource Allocation in the Health Service*. Occasional Papers in Social Administration No 81. London: Bedford Square Press.

NHS (1994a). *Improving NHS Dentistry*. Cm 2625. London: HMSO.

NHS (1994b). *NHS Executive. Revenue Allocations 1995/96*. FDL(94)68. Leeds.

OECD (1993). *OECD Health Sytems: Facts and Trends 1960–1991*. Paris: OECD.

OECD (1994). *OECD Economic Surveys 1993–1994: United Kingdom*. Paris: OECD.
PPAC (Prospective Payment Assessment Commission) (1992). *Medicare and the American Health Care System. Report to the Congress*. Washington DC: PPAC.
PPRC (Physician Payment Review Commission) (1993). *Fee Update and Medicare Volume Performance Standards for 1994*. Washington DC: PPRC.
Royal Commission on the NHS (1979). *Report*. London: HMSO (Cmnd 7615).

Part 3

Institutional Issues

Chapter 12

German Sickness Funds under Fixed Budgets

Sabine Richard
Karl-Heinz Schönbach

INTRODUCTION

The German health insurance system is facing severe changes in the immediate future. The Gesundheitsstrukturgesetz (Health Care Structure Act – GSG) which came into force in 1993 has introduced several competitive elements into the system. In this chapter we will outline the basic changes and their effects on the sickness funds.

The GSG has introduced two important elements. The first element, designed for short-run cost containment, are collective fixed budgets for the different sectors in the health care system. This includes fixed budgets in ambulatory care on the *Länder* level, in dental care on the *Länder* level, for every hospital individually, and for pharmaceuticals, remedies and devices. These budgets and their effects are discussed in detail in other chapters.

The second element, designed for the long-run competitive restructuring of the health care system, is the risk compensation scheme and the introduction of free choice for the insured persons among the sickness funds on the insurers' side and the reform of the hospital payment system introducing prospective rates for certain procedures on the side of the hospitals.

In this chapter we focus on this competitive restructuring, especially the legal regulations and questions left open by the legislator. We also discuss strategies proposed by the sickness funds to overcome the dilemma of both legally induced competition and uniform regulations concerning the coverage, financing and provision of services.

COMPETITIVE RESTRUCTURING – LEGAL REGULATIONS AND OPEN QUESTIONS

The GSG has brought about significant changes of the role of the statutory sickness

Fixing Health Budgets: Experience from Europe and North America.
Edited by F. W. Schwartz, H. Glennerster and R. B. Saltman. Technical editor R. Busse.

funds. Competition among the sickness funds is expected to enhance efficiency and customer orientation. Two important preconditions for effective competition introduced by the GSG are:

- a risk compensation scheme to equalize the different risk structures
- the freedom to choose among different sickness funds.

These two elements were introduced in two steps, with the risk compensation scheme effective since 1 January 1994 and the free choice effective since 1 January 1996.

Risk Compensation Scheme

The traditional segregated system of sickness funds consists of two major types of funds. The primary funds (*Allgemeine Ortskrankenkassen, Betriebskrankenkassen, Innungskrankenkassen*) cater mainly for blue-collar and white-collar workers, whereas the substitute funds (*Ersatzkassen*) cater mainly for white-collar workers. In addition, special funds exist for farmers, miners and sailors. Therefore, both health risks (in terms of income, education, age and occupational hazards) and health services utilization and expenditure are spread unevenly among the different types of sickness funds. Until the end of 1993, only the expenditures for retired persons were shared equally among all sickness funds.

Since 1994, factors which determine revenues and expenditures of a sickness fund are accounted for in the risk compensation scheme (*Risikostrukturausgleich*) established by the GSG. On the revenue side the sickness funds are compensated for the income level of their insured members. As the premium paid by the member is calculated as a percentage of his or her labour income, the compensation for the income prevents the sickness funds from selecting mainly high income persons for membership. On the expenditure side the sickness funds are compensated for the factors age, sex, number of dependent family members (who are covered free of charge by the statutory sickness funds) of the members. These factors are chosen as indicators for the expected expenditures. To calculate the compensation level of a sickness fund the members are grouped according to sex (two groups) and age (15 groups) in 30 categories. For each category an average expenditure profile has been determined on a nationwide level. The risk compensation scheme, therefore, induces the sickness fund to control its expenditures for age/sex groups with above average expenditures and to investigate the causes.

The risk compensation scheme is supposed to equalize the morbidity risks of the insured persons which the sickness fund can only marginally influence. In the past the uneven distribution of morbidity risks due to restricted access to the different sickness funds for white-collar and blue-collar workers has led to large differences in contribution rates. These have been equalized since the introduction of the risk compensation scheme. A large reallocation of funds has taken place with the local and the guild sickness funds being the winners, and the substitute funds and the company based sickness funds being the losers.

Nevertheless, there still are some shortcomings, because age and sex are only weak indicators for the expected expenditures of one member. Moreover, there are large regional variations in health care expenditures, which cannot be explained by the factors considered.

Currently, the regionally operating local sickness funds are arguing in favour of the introduction of further compensation factors particularly with regard to the high regional variations in expenditures presently not covered by the risk compensation scheme. But from a system point of view the regionalization of the competition between sickness funds tends to conserve the unexplained variations. As the contribution rates of all sickness funds in high expenditure regions would then be high, the competitive pressure to look for the reasons of large regional variations would be severely weakened. Assuming that a large part of the variation can be attributed to the density of physicians and hospitals (under the existing reimbursement regime the number of services produced increases with the number of providers), in a regionalized competition the insured persons and their employers in regions with high expenditures would be held liable for the regional overcapacity by paying high contribution rates.

Free Choice of Sickness Fund

There are two sides of the regulations permitting the free choice of a sickness fund: on the side of the sickness funds, the general regional funds (*Allgemeine Ortskrankenkassen*, AOK) and the substitute funds have the obligation to contract with every customer who has chosen to become a member. The company-based sickness funds (*Betriebskrankenkassen*, BKK) and the guild sickness funds (*Innungskrankenkassen*, IKK) are allowed to choose whether to open themselves up to a new clientele. If they decide to do so, they have the obligation to contract, as well. By law, the special funds for farmers, miners and sailors will remain "closed". Before the competitive market for members, the market share of the different groups was: AOK 46%, substitute funds 33%, BKK 12%, IKK 5%, miners fund 3%, farmers funds 2%, sailors fund 1%.

On the members side, every person – apart from farmers, miners and sailors – now has the choice among about 15 substitute funds, one general regional sickness fund, in certain professions one guild sickness fund, or, if existing, a company-based sickness fund.

It is interesting to note that the regulation preventing persons who have opted out of the statutory health insurance scheme to become a member of a private insurance from re-entering the statutory health insurance scheme will remain in force.

Open Questions

In the following sections the opportunities for competitive behaviour given to the sickness funds by the legislator are described. The legislator has hitherto failed to put in concrete terms, how competition among the sickness funds should be structured.

Contribution Rates

As the statutory health insurance has not been competitive in the past, it is not known yet what role the contribution rates will play in the decision of the potential sickness fund clients. A low contribution rate relative to the competitors' might arouse the suspicion that the sickness fund is purchasing low-quality health care services or offers a low coverage. It is still unknown how large the price elasticity of the clients will be and how large the segment for cheap sickness funds will be in consequence. Studies conducted in the past have shown that most of the insured persons are neither aware of the contribution rate they pay each month, nor the share paid by their employer. In the past, contributions rates had almost no role in the strategies of the sickness funds, because most of the sickness funds had a very limited clientele entitled for membership. Only white-collar workers had a choice between several sickness funds. Most blue-collar workers were restricted to membership in the local sickness fund. They had to accept a steady increase in their contributions without being able to leave. Moreover, the steady rise in contribution rates was common to all sickness funds.

Opposition to the expansive tendencies of the health care sector was not primarily brought forward by the insured persons, but by their employers who have to pay half of the bill. High ancillary wage costs are also gaining importance in the discussion about the international competitiveness of German industries.

It is irrational to think that the low price elasticity of demand observed in the past will hold under competitive circumstances. The employers, at least, will have an incentive to direct their employees to the cheapest sickness funds, or will lobby to have their share of the contribution rate fixed at the average level.

A further and probably more important question is how the sickness funds will be able to influence the contribution rates at all. A first option is to reduce administration costs by an efficient management of the sickness funds. But the administration costs of an average sickness fund account for only for 5% of its expenditures. This figure shows the efficiency of the German system of statutory sickness funds in international comparison. Administration costs of the American private insurers, for example, total 11.9%. German private insurers face even higher administration costs on average: 12.7% (Schwartz and Busse, 1994). Therefore, the sickness funds have to look for more effective fields for their competitive activities.

The Range of Services Covered

Currently, about 95% of the health care expenditures cannot be influenced by individual product policies of the individual sickness fund because of legal restrictions. Product differentiation, therefore, can take place only in the 5% margin. It is expected that the share of services a sickness fund offers to its clients on a discretionary basis will rise in the future. The only legal way to expand the range of services beyond the legally determined catalogue is to offer additional services in the field of health promotion, maintenance and prevention. As the wording of the legal prescription is very vague, sickness funds have initiated all kinds of courses and events to promote healthy behaviour, which are undertaken primarily as mere marketing devices. The effect on the behaviour of the participants in terms of changed health habits is rarely evaluated (Epidemiologische Forschung Berlin, 1994).

Competition among the sickness funds is very likely to lead to a broader range of covered services, as the client preferences, competition will disclose, currently consist of a criticism of orthodox medicine and a desire for alternative healing methods. A recent opinion poll indicated 81% of the Germans value alternative methods positively and 60% asked for inclusion of the services of non-medical practitioners in the social health insurance system (*Die Woche*, 1994). Once one fund includes the large-scale coverage of natural medicine, the competitors will have to follow, although the effectiveness is doubtful and the financial consequences are unforeseeable. The financial costs arising from natural healing methods will cause contribution rates to rise, unless they can be used to substitute the traditional methods of orthodox medicine.

On the contribution side of the market, the decision makers in our health care system are – based on past experience – unwilling to trust the market mechanism to solve the problem of inflated coverage in the long term. The different cost containment laws of the 1980s, especially the Health Care Reform Act (Gesundheitsreformgesetz, GRG), restricted the sickness funds (individually and with regard to the different kinds of sickness funds) more and more in their negotiating scope: not only in the matter of negotiations with the organizations of health care service providers, but also concerning the range of services that can be offered to clients. The GRG almost entirely abolished the free scope of regulations that the sickness funds could define in their statutes before. As already mentioned, health promotion is the only exception (SGB V, §20), but even there rules put forward by the supervisory board for competition between sickness funds (*Bundesversicherungsamt*) limit the scope for the further introduction of health promoting activities and restrict marketing expenditure (*Gemeinsame Wettbewerbsgrundsätze der Aufsichtsbehörden der gesetzlichen Krankenversicherung*, 1994). This indicates a scepticism over the workability of competitive structures. In addition, the *GSG*'s introduction of fixed budgets in all sectors of health care delivery with the intention to prevent dramatic cost increases in health care provision restricts severely the ability of the sickness funds to compete through product differentiation in terms of services offered.

Selection of Clients

We have seen that the array of competitive strategies at the disposal of the sickness funds is very limited. Another strategy, which can profitably be pursued by a sickness fund but which does not lead to workable competition, is "cream skimming". Compensation for certain risk factors including sex and age is carried out, but there is ample scope for risk selection. The risk factors included in the risk compensation scheme can – according to international studies – explain only up to 3% of the variation in *individual* health expenditure (van de Ven and van Vliet, 1992). Therefore, a sickness fund has an incentive to address primarily those people who are likely to cause lower expenditures than the underlying standard expenditures in their risk group.

The question is, how can sickness funds detect the "good risks"? As the variation of services covered and organizational structures is small between the sickness funds, the scope for risk selection is restricted as well, but the sickness funds have nevertheless several means at their disposal. The first means is active repelling of unwanted clients, in spite of the legal obligation to contract with any person. Repelling strategies include impeding access to their outlets for handicapped or aged persons or disseminating information and advice selectively. These strategies are illegal and the sickness fund using these methods too extensively is likely to arouse public criticism and to lose its good reputation in the sensitive health care sector, where commercial practices are observed very critically by the public. A more subtle and elegant strategy for selecting risks is to tailor products to the needs and preferences of healthy people. Target groups are wealthy and well-educated persons.

Another method of selecting healthy persons is a rebate of contributions in case of low personal health expenditures as is practised widely in the private health insurance market. Currently, the rebate of premiums is allowed only on an experimental level. Sickness funds can test it for five years in order to reduce moral hazard behaviour of their members. As first results from the experiment carried out by several company-based sickness funds suggest, the effect of the rebate on the behaviour of the members is relatively weak (Malin and Schmidt, 1995). The underlying intention of sickness funds planning to introduce the rebate of premiums in 1996 is not primarily reducing expenditures of their members but to improve the risk structure within the sickness fund. The rebate of premiums attracts persons who expect that their personal health expenditures in the near future will be low. With their decision to choose a sickness fund offering premium rebates for low expenditures these persons convey valuable information about their health status to the sickness fund. Other sickness funds will follow until they find a better selection strategy.

As some of the selection strategies are expensive (high administration costs) and of no use when carried out by all competitors, it might be economically rational to abstain from them collectively. What is needed is a well-designed framework for the competitive process in the German health care sector.

Contracting with Providers

Another central element of the current transitory situation is that the provider side lacks competitive elements. As an important example the contractual relations to physicians in private practices are described below (Figure 12.1)

The sickness funds currently are not allowed to negotiate directly with single physicians. Contracts are made between the regional federations of each type of sickness fund and the regional physicians' association. These parties negotiate a prospective budget for all physicians, which is distributed by the regional physicians' association, more or less according to the fee-for-service principle. The sickness funds can only marginally influence the distribution of the budget among the physicians.

In contrast to most other economic sectors of the German economic system, the physicians' sector is characterized by self-regulation. Like other professions, the medical profession has achieved the exclusion of any competitive elements, to date. Physicians are not allowed to reveal any specialization beyond the fields defined by the professional chambers. But the revelation of qualifications acquired in the past cannot be taken as a good indicator of quality in the present. Consumerism in medical affairs is so far unknown in Germany. Neither the sickness funds nor the patients have sufficient information to select a physician rationally (Richard, 1994). Nevertheless, competitive behaviour among the physicians may be observed, because the number of sickness fund-affiliated physicians has increased dramatically without similar increase in the number of patients. More and more, the physicians regard other physicians as rivals rather than as colleagues (Andersen and von der Schulenburg, 1990). As open competition is prohibited by professional rules, physicians in the ambulatory sector use inefficient and expensive signals of their quality and

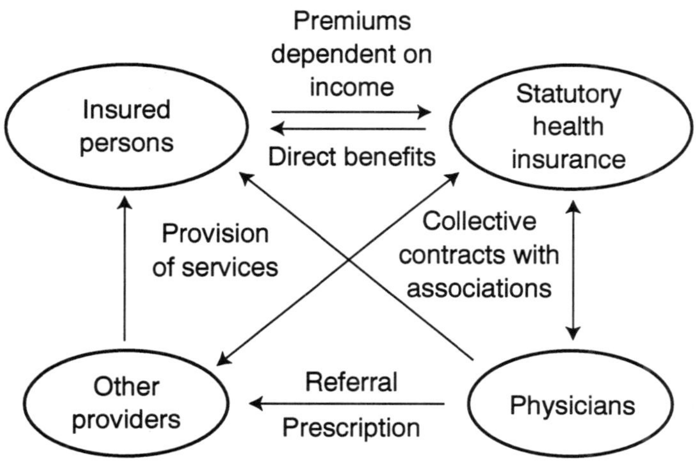

Figure 12.1 Current contractual relations in statutory health insurance

competence to attract patients, e.g. high technology equipment in their practice or generous prescribing habits (Richard, 1993).

Quality assurance is performed by the profession, but the engagement the profession has shown in the past has not been convincing (Selbmann et al., 1994). Up to now the profession has claimed that the independent single practitioner, ideally paid for by a fee-for-service system is the leading organizational form of health care practice, ignoring that medical and technical progress call for co-operation and integration to make use of efficiency gains by joint use of technical appliances and practice staff. Only recently, the physicians' associations have recognized the necessity for more inter- and intra-professional co-operation in the health system. In 1995, the professional code has been be adapted to allow the co-operation between physicians and paramedical health care providers in one practice.

There is no scope for the sickness funds to negotiate conditions different from the single practice/fee-for-service structure. The only field where the sickness funds are legally allowed to negotiate with single providers is in guarantees for the quality of oral health care.

The rigid legal structure has two implications. On the one hand, the sickness funds are denied an important field of competitive action, because they cannot influence providers by setting financial incentives or offer their clients different forms of health care service provision. This means, that an important function of competition – to give rise to organizational innovations reflecting changing needs of the patients – cannot be realized.

On the other hand, the supply side of the market – especially the market for physicians' services – remains monopolized. The competing sickness funds as purchasers of health services of their insured persons on the one side of the market are facing a highly monopolized structure on the provider side of the market. Being monopolists, the regional physicians' associations resist any organizational restructuring of the provision of medical care. As representatives of all physicians in one area – i.e. physicians of all specialities and quality levels – the regional physicians' associations have to pursue a strategy that satisfies the economic needs of every member. Therefore, a sickness fund wishing to select providers according to individually contracted quality criteria, threatens the intra-professional balance of power.

The only competitive element which remains is the size of the collective budget negotiated between each type of sickness fund and the regional physicians' association or the remuneration of additional services outside the budget. The substitute funds especially have used their good market position in the past to pay providers generously hoping that the physicians will use their influence on patients in their favour. But the legal determination of budget increases and the introduction of the risk compensation scheme has inhibited the use of this instrument. Moreover, the patients are the group which has to be addressed in competition by the sickness funds. Their gain from the above mentioned strategies was only very small.

As was mentioned earlier, only some people previously had the right to choose their sickness fund. Their experiences under this previous system have led many decision makers to believe that competitive contracting with providers can only lead to increased costs. This ignores the fact that – depending on the price elasticity of the clients – the previous strategy (of expanding services) might not remain profitable under competitive circumstances, and might, therefore, be changed in practice. Moreover, the physicians' associations are objecting heavily to the sickness funds' demand for more competition on the supply side. The federal associations of sickness funds has urged the legislator to liberalize the existing contract regulation within the next legislative period.

STRATEGIES OF THE SICKNESS FUNDS

The changing structure of the sickness funds calls for a similar restructuring on the provider side and a redesign of traditional contracting relations. Despite the regulators, desire to stimulate competition in the health care sector, the simultaneous desire to control expenditure means that the only major area for competition left is in the area of innovative service delivery. The federal associations of the sickness funds have, therefore, reached unanimity in demanding a new legal setting for competitive strategies. In the view of the sickness funds, deregulation, widely promoted in many sectors of the German economy by the German federal government, is desperately needed with regard to the physicians' profession. Therefore, they demand more flexibility in the contractual relations between sickness funds and providers. Both the federal associations of general regional sickness funds (AOK) and of company-based sickness funds (BKK) propose the integration of managed care elements in the German health care system (Enthoven and Kronick, 1989; Enthoven, 1994; van de Ven, Schut and Rutten, 1994).

What is managed care? Although there is no single standard definition, the American Medical Association defines managed care as "the control of access to and limitation on physician and patient utilization of services by public or private payers or their agents through the use of prior and concurrent review for approval of or referral to service or site of service, and financial incentives or penalties" (Iglehart, 1992). To achieve these aims, outcome or performance measures must be included. The main objective is to improve efficiency, which in this context means the maximization of health outcomes for a given cost. Managed care in this respect is an approach in which the traditionally divided responsibilities of providing care and providing payment are united (Hall, 1994; Baumberger, 1995). A central element of most managed care systems is that care is prepaid. Prospective budgets calculated on the basis of the number of treated persons modified by standard morbidity characteristics are common.

When comparing the above mentioned characteristics of managed care with the current German system, the fundamental differences become evident. The provision of health care services in Germany lacks any co-ordinating activities. Any measures that would interfere with medical decision making have been rejected by a medical profession careful to protect its members' therapeutic freedom.

The financial and organizational segregation of health care according to different sectors has led to several shortages in the provision of health care services (Bundesministerium für Gesundheit, 12/8570, 1991), for example:

- unspecified referrals to specialists or to hospitals
- repeated diagnostic procedures in the different sectors
- under-use of ambulatory surgery
- insufficient co-operation of physicians with other physicians, the hospital sector, and the paramedical professions
- over-capacity in medical technology
- lack of evaluated quality assurance programmes.

International experience with managed care has shown that medical care is much more accessible to standardization and external quality assurance than the medical profession has, for many years, maintained (Grimshaw and Russell, 1993).

Not only has medical treatment itself come under scrutiny, but also the organizational and financial framework, in which it takes place has been subject to intensive discussion in the last years. In the international literature several comparative studies which analyse the performance of different forms of care provision may be found (Safran, Tarlov and Rogers, 1994; Miller and Luft, 1994). Several providers of health care were compared according to the health status of patients treated, cost efficiency, and patient satisfaction. In the past outcome indicators such as these have not been of any interest in Germany. The competitive restructuring of the social sickness funds' has already directed the sickness funds' interest towards consumer satisfaction. When the sickness funds are given more responsibility for their contract relations with providers, the interest in outcome control will probably increase.

With the experience of global budgeting gained in Germany over the last two years, and the positive experiences with managed care in other countries, the Federal Association of Company-based Sickness Funds (Bundesverband der Betriebskrankenkassen – BKK) is considering the combination of both measures. The BKK has developed a strategy to implement elements of managed care in the German health care system.

The focal point of the programme is the central role of the physician in ambulatory health care. In Germany, the physician decides on several times his or her own fee without bearing economic responsibility for the efficiency of the triggered cost. It is necessary therefore, to establish a different financial framework within which the physicians may act. The BKK's aim is to combine medical decisions and financial responsibility without leaving the physician with the complete morbidity risk of the patients. Therefore, we intend to break up the strict financial and organizational segregation of the different sectors in health care and to establish a system of communicating budgets beneath the level of the global budgets. *Combined budgets* will be a financial framework for competition among the sickness funds as purchasers of medical services on the one side of the market and between the providers on the other side.

The BKK intends to establish or to contract with a provider network, either a group of physicians and other providers in an integrated health care centre or a network of affiliated group practices co-operating intensively. The range of services provided would comprise of preventive activities, curative care, ambulatory services, rehabilitation and co-ordination of care. This corresponds to the range of services already covered now. The decisive difference to the existing system is not the range of service, but the structure of provision. The BKK is aiming at a system of financial and other motivational incentives, to create an environment where physicians and other providers of health care services, can base their decisions on medical categories and guidelines. A central element is professional management. Co-operation among providers in the network guarantees co-ordinated treatment of patients. Referrals to hospitals require a second opinion. (See Figure 12.2.)

The clients of a company-based sickness fund have the choice between the traditional system and the special health care plan. If they choose the special health care plan, they commit themselves for a certain period of time to abstain from directly consulting providers not integrated in the network. The decision of the patients to participate in the managed care system or not will reflect their satisfaction with the services offered and will serve as a signal to the providers over the need to enhance the quality of their services and to the sickness fund over the need to look for alternative providers.

The provider network is entitled to a prospective budget. This budget is calculated on the basis of the standard expenditure profiles underlying the risk compensation scheme. The budget, therefore, comprises the expenditures for all health care services and gives the participating physician or the group of physicians the financial responsibility for co-ordination of the whole treatment, not only for the services produced by him or herself but also the services they induce and delegate (Figures 12.3 and 12.4).

Co-ordination of treatment and supervision of the budget require professional management in the network. "Managed care" is a novelty in the German health system, because the unique financial structure of the German health insurance system has hitherto prevented any realization of managed care. Together with the physicians' organizations the BKK is, therefore, developing a quality assurance system adapted

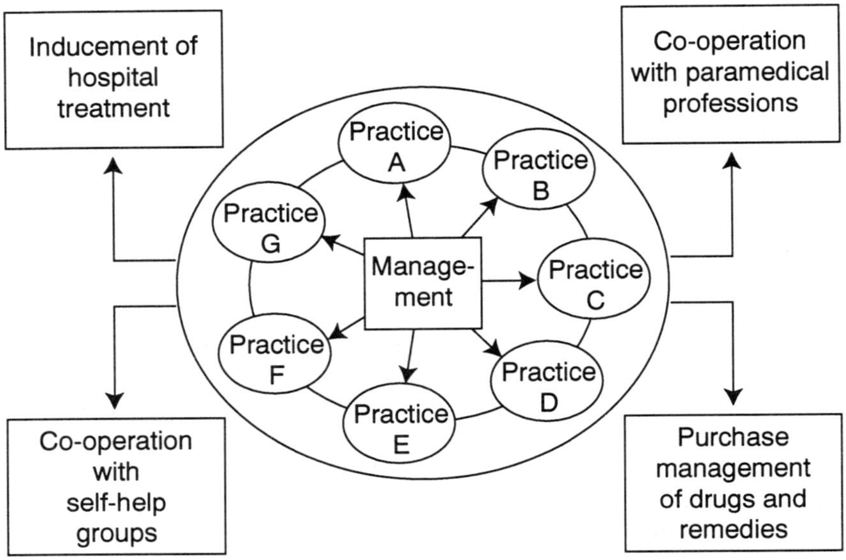

Figure 12.2 Proposal: Provision of medical care

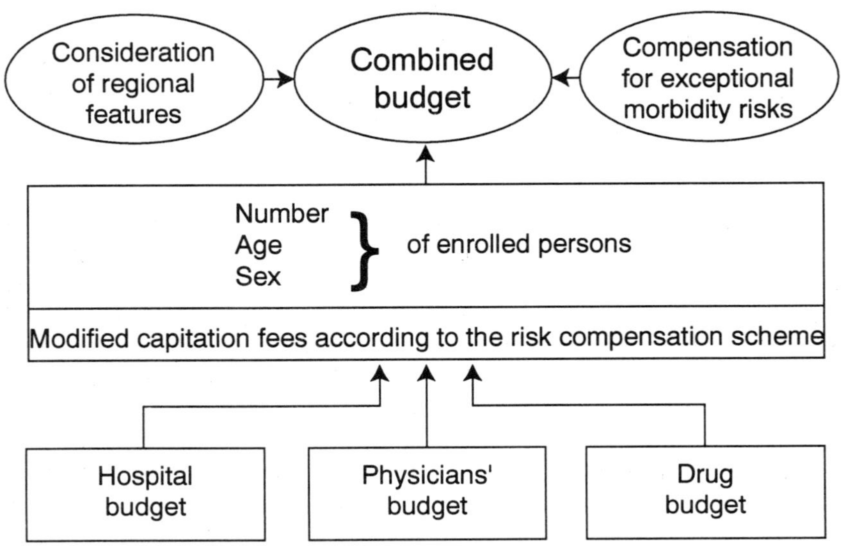

Figure 12.3 Proposal: Combined budget

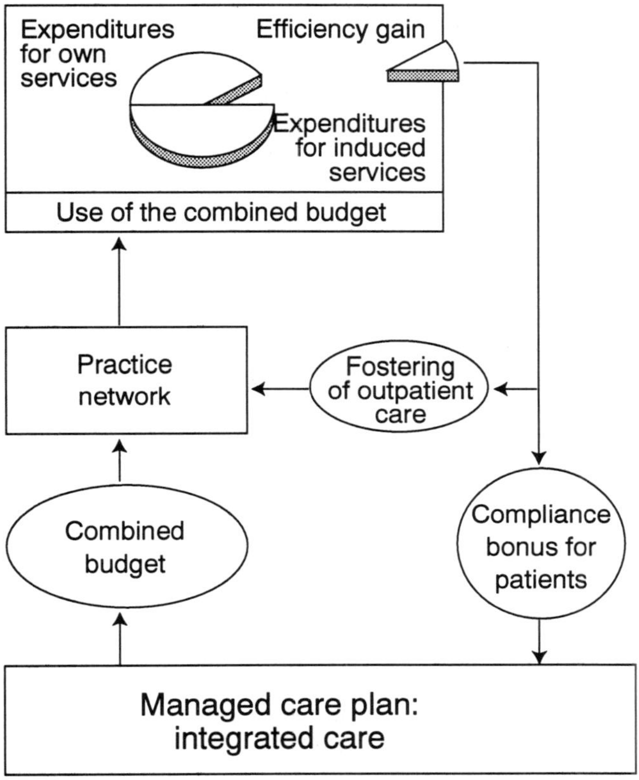

Figure 12.4 Proposal: Combined budget

to the special needs of an integrated provider network paid for by a prospective budget. The BKK expects a quality assurance scheme to comprise the following elements.

First, the physicians participating in the network are required to meet certain qualification criteria, e.g. postgraduate specialization, knowledge in psychosomatic care, clinical epidemiology, experience with electronic expert systems, experience with co-operation (previous work in group practices) and previous participation in quality circles.

Second, the organizational structure is based on transparency, communication between the providers, and the development of practice guidelines. Transparency will be achieved by internal computer networks and by comprehensive registration and analysis of treatment schemes. Communication will be established by routine case conferences and a system of second opinion. Electronic processing and analysis of own data and international studies will allow the development of standards of medical practice including a restricted catalogue of recommended pharmaceuticals. Another element of quality assurance is the remuneration of the providers, which should give the physicians the incentive to adhere to the developed standards.

Third, the BKK expects the development of indicators which reflect outcome quality. Apart from the medical quality, the satisfaction of the BKK's insured members will be an important element.

Combined budgets and integrated health care centres represent the German adaptation to managed care. Once the legal preconditions are set, the company-based sickness funds are ready to negotiate with teams of providers who offer comprehensive health care according to jointly defined standards and commit themselves to serious quality assurance. The BKK are not aiming at bureaucratic and hierarchical institutions but rather expect processes of self-organization on the side of the providers.

REFERENCES

Andersen, H.H. and von der Schulenburg, J.M. (1990). *Konkurrenz und Kollegialität: Ärzte im Wettbewerb*. Berlin: edition sigma.

Baumberger, J. (1995). Managed care – a US insanity or future European reality? In: M. Lonsert, K.J. Preuß, and E. Kucher (Eds), *Handbuch Pharma-Management*. Wiesbaden: Gabler, pp. 153–82.

Bundesministerium für Gesundheit (1991). Zweiter Bericht des Bundesministeriums für Gesundheit zur Entwicklung der Beitragssätze in der gesetzlichen Krankenversicherung und zur Umsetzung der Empfehlungen und Vorschläge der Konzertierten Aktion zu Erhöhung der Leistungsfähigkeit, Wirksamkeit und Wirtschaftlichkeit im Gesundheitswesen. *Bundestagsdrucksache*, 12/8570, pp. 10–88.

Enthoven, A. (1994). On the ideal market structure for third-party purchasing in health care. *Social Science and Medicine*, **39**, 1413–24.

Enthoven, A. and Kronick, R. (1989). A consumer-choice health plan for the 1990s (two parts). *New England Journal of Medicine*, **320**, 29–37 and 94–101.

Epidemiologische Forschung Berlin (1994). Untersuchung zur Umsetzung des §20 SGB V durch die Krankenkassen. Berlin.

Grimshaw, J.M. and Russell, I.T. (1993). Effect of guidelines on medical practice: a systematic review of rigoruos evaluations. *The Lancet*, **342**, 1317–22.

Hall, M. (1994). Managed care in Europe – myth or reality. *SCRIP Magazine*, October 1994, pp. 32–4.

Iglehart, J.K. (1992). Health policy report: the American health care system: Introduction. *New England Journal of Medicine*, **326**, 962–7.

Malin, E.-M. and Schmidt, E.M. (1995). Beitragsrückzahlung in der Gesetzlichen Krankenversicherung. *Arbeit und Sozialpolitik*, No. 1–2, 55–68.

Miller, R.H. and Luft, H.S. (1994). Managed care plan performance since 1980. *JAMA*, **271**, 1512–19.

Richard, S. (1993). Qualitätssicherung und technologischer Wandel im Gesundheitswesen. Baden-Baden: Nomos.

Richard, S. (1994). Wettbewerb auf dem Markt für Arztleistungen. *Arbeit und Sozialpolitik*, No. 3–4, 35–40.

Safran, D.G, Tarlov, A.R. and Rogers, W.H. (1994). Primary care performance in fee-for-service and prepaid health care systems. *JAMA*, **271**, 1579–86.

Schwartz, F.W. and Busse, R. (1994). Das Wachstum in der Medizin – Mythen und Fakten. *Arbeit und Sozialpolitik*, Nos 9–10, 10–17.

Selbmann, H.K., Pietsch-Breitfeld, G., Blumenstock, G. and Geraedts, M. (1994). Evaluation der Qualitätssicherungs-Maßnahmen im Gesundheitswesen. In: Bundesministerium für Gesundheit (Ed.), *Maßnahmen der Medizinischen Qualitätssicherung in der Bundesrepublik Deutschland – Bestandsaufnahme*. Baden-Baden: Nomos, pp. II 1–II 21.

van de Ven, W.P.M.M., Schut, F.T. and Rutten, F.H.F. (1994). Editorial: Reforming the market for third-party purchasing for health care. *Social Science and Medicine*, **39**, 1405–12.

van Vliet, R.C.J.A. and van de Ven, W.P.M.M. (1992). How can we prevent cream skimmng in a competitive health insurance market? The great challenge for the 90s. In: P. Zweifel and H.E. Frech (Eds), *Health Economics Worldwide*. Dordrecht: Kluwer, pp. 23–46.

Die Woche (1994). Forsa-Umfrage for the weekly magazine Die Woche, as reported in: *Gesundheitspolitische Umschau*, November 1994.

Chapter 13

The Possible Effects of a New Hospital Financing System in Austria

Karl P. Pfeiffer

INTRODUCTION

The Austrian hospital system has a very long history, which is also reflected in a complex but not up-to-date financing system. Almost 100% of the expenses are covered by public sources and only a very small amount by private ones, mainly from private insurance companies.

The hospital financing system is a dual one. In principle, the hospital and its equipment are financed by the owner of the hospital, the province and local government as well as some orders, with some support from the Ministry of Health. The operating costs are paid from different sources: more than 50% comes from public social insurance companies based on the length of stay and independent from the diagnosis or from special medical procedures. The Ministry of Health also has a special fund (*Kranken-anstaltenzusammenarbeitsfonds*) that gives lump-sum subsidies to owners of hospitals, dependent on the number of beds and special functions, and on the deficit. The difference between operating costs and income from the social insurance companies and the fund of the Ministry of Health (= deficit) has to be paid by the owner. In the actual system the hospital itself makes no losses and there are no sanctions. In principle the Ministry of Health has no direct influence on investments and operating costs, it has only a co-ordinating function and can make policy statements.

Because of a missing general responsibility the Austrian hospital financing system has no incentives for cost reduction and no instruments for cost control. Moreover the actual system has inverse incentives by keeping patients longer than sometimes necessary in the hospital, because basic income for the hospital depends on the number of inpatient-days. The average costs per inpatient-day differ by a factor 4 to 5 within Austria. This difference is of course due to different specialities (basic care, academic hospitals) but there seems to be also a big difference in efficiency.

Fixing Health Budgets: Experience from Europe and North America.
Edited by F. W. Schwartz, H. Glennerster and R. Saltman. Technical editor R. Busse.

Since this large increase of the costs of hospitals – as in most other countries – new regulatory mechanisms have to be developed. The driving force behind this effort is the Ministry of Health. This wants to establish not only a pure financial mechanism, but also to consider the quality of care as well as special functions within the Austrian hospital system. Referring to these developments a planned market for the Austrian hospital system should be established.

This chapter will demonstrate the steps towards a new hospital financing system for inpatients in Austria and will discuss its possible impact on the hospitals.

THE AUSTRIAN HOSPITAL SYSTEM

Austria has a very highly developed medical system and everybody has almost free access to this system, because of a well-developed public social insurance system. Funds are coming from health insurance fees as well as from taxes. Furthermore, for each inpatient-day a patient has to pay about $5.50 (for a maximum of 28 days a year), with some exceptions for people with low income. Some people also have private health insurance, which covers a higher hotel quality and free choice of doctors.

Some basic data of the Austrian hospital system can be found in Table 13.1 (Österreichisches Statistisches Zentralamt, 1992). Compared to many other European countries the rate of hospital admission is rather high and the average length of stay is long. About 50% out of the 323 hospitals with approximately 80% of all hospital beds get financial support from the government.

Table 13.1 Some basic data of the Austrian hospital system for 1991

Population of Austria	7 795 786
Number of hospitals	323
Number of hospital beds	80 476
Hospital days	22 959 446
Number of inpatients	1 787 092
Number of inpatients per 1000 inhabitants	229
Average length of stay (days)	12.8
Average costs per in-patient (in $)	2 628
Average costs per hospital day (in $)	205
Total costs for inpatients (in $)	4 697 217 818
Number of ambulatory cases in hospitals	3 983 085
Costs for ambulatory care in hospitals (in $)	658 088 029
Total hospital costs (in $)	6 039 237 193

Note: $1 = 10 AS

THE MODEL FOR A HOSPITAL FINANCING SYSTEM IN AUSTRIA

Since 1989, in all Austrian hospitals, which get support from the fund of the Ministry of Health, a minimum basic data set (MBDS) for all inpatients has to be documented. This data set consists of the main diagnosis, up to nine additional diagnoses (ICD-9), medical procedures according to a special catalogue, number of each procedure, type of admission and discharge, date of admission and discharge, length of stay, age, gender, code of departments (which is especially important for intensive care units), postal code, social insurance company, etc. The main and additional diagnosis should be coded according to the ICD-9 with four digits. The catalogue of procedures contains approximately 800 mainly expensive and some very frequently applied surgical and non-surgical procedures. The MBDS is the data base for the development of the new hospital financing system.

In order to identify homogenous groups of patients in 20 Austrian hospitals in 1991 and 1993 for approximately 600 000 patients, the costs per patient has been computed in each hospital. This data set can be considered as a learning sample. In a first step from a pure medical viewpoint this data set has been partitioned into a "surgical group" and a "non-surgical" group by a group of medical experts. Furthermore, the surgical group has been divided into smaller subgroups with similar surgical procedures and the non-surgical group has been divided into subgroups with similar main diagnosis. In summary there are about 240 "main groups". A regression tree algorithm (Figure 13.1) has been used to identify homogenous subgroups within the "main groups" (Pfeiffer, 1991, 1994). As split criteria, that means as predictors, the variables of the MBDS and as dependent variable the costs for each patient have been used. Additionally further split criteria such as additional diagnosis yes/no, medical procedure yes/no, number of medical procedures, age classes, etc., have been defined. To identify plausible subgroups also from a medical viewpoint a hierarchy of the split criteria can be defined. This hierarchy means that the selection of the split criteria is not only due to the optimization criterion but it depends also on a pre-defined hierarchy level.

The idea behind this classification system is very similar to the idea of the DRGs. The aim of this new financing system is to take into account the diagnosis, selected surgical and non-surgical procedures and other items of the MBDS as an alternative to the system now, which is based on the length of stay.

Paying attention to the special medical culture in Austria statistical classification procedures have been used to identify homogenous groups of patients. "Homogenous" means that the groups should be homogenous in a medical sense and that patients of one group consume a similar amount of resources. As a global statistical goodness of fit criterion for a single regression tree the median absolute distance (MAD) was used:

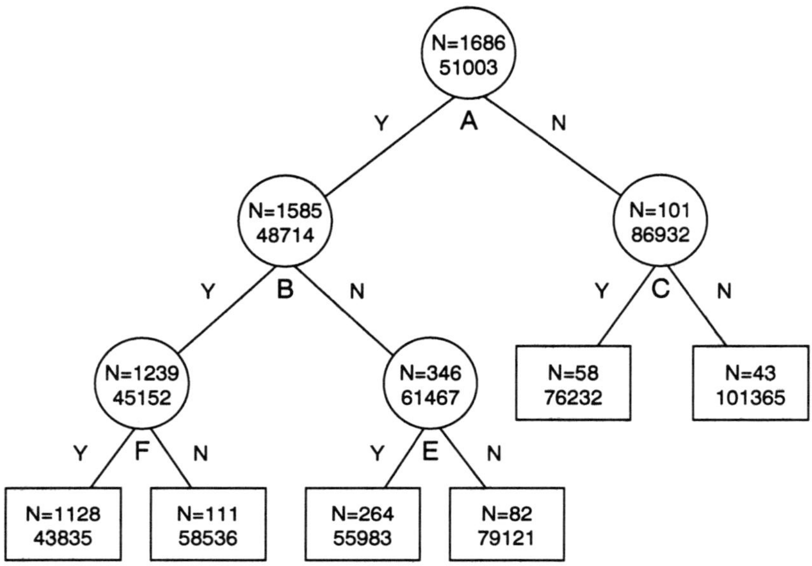

Figure 13.1 Example of a regression tree for the data set *ojm156*; sample sizes (N) and mean values of the nodes; the split criteria are indicated by A, B, C, D, E, F

$$MAD = \sum_{i=1}^{k} \sum_{j=1}^{n_i} | y_{ij} - d(X_j)_i | \tag{13.1}$$

y_{ij} ... dependent variable in node i for case j

$d(X_j)_i$ ... predicted value in node i for case j using the independent variables X_j of case j

Furthermore, a stopping rule, which is the minimal number of observations in one node, has to be defined. For each local split the variable is selected which maximizes a distance function between the two resulting groups and which has the highest level in the hierarchy of variables.

After evaluation of the results of the statistical classification procedure by different medical specialists about 1500 different subgroups have been defined. Because of the sometimes strong skewness of the distribution of the dependent variable, namely the costs, the median of each subgroup is used as a characteristic parameter for each subgroup. This characteristic parameter of one node is called LDF-points (*Leistungsorientierte Diagnose-Fallgruppen-Punkte*). They are similar to the DRG-points. For each subgroup the average length of stay is computed and an 80%-confidence interval for the length of stay.

Figure 13.1 shows an example of a regression tree for one main group. As can be seen from the mean values of the terminal nodes, the lowest mean value is 43 835 and the highest one is 101 365. The big range of the mean values of terminal nodes indicates the importance of the split. Based on the MBDS each patient now can be classified exactly into one subgroup and the LDF-points can be assigned. Because of the strong dependency of this system on the diagnoses and on medical procedures this system cannot be used for long-term patients.

Comparison of Actual Hospital Costs with the Results of the New Financing Model

All Austrian hospitals have to report a calculation of their costs on department level. These data include information about the number of beds, number of medical doctors, nurses, etc., and also about the running costs. Using the sum of LDF-points for each department and in summary for each hospital the theoretical income with the new financing system for one year can be computed and it can be compared with the costs of this year. This calculation has been done for the years 1991 and 1993.

To get the value of a LDF-point the sum of the costs for the hospitals under consideration is divided by the sum of LDF-points for these hospitals. This means that a fixed amount is available for the hospitals under consideration and this amount is distributed to the hospital according to their case mix. The aim of this analysis was to study the reproducibility of the costs by the new system for acute hospitals on one side and the possible financial effects of the new system for the owners on the other side. A correlation coefficient between reported costs and calculated costs for acute hospitals of $r > 0.9$ was found. But for some hospitals the differences are rather big and explanations of the residuals have to be found. These comparisons of the running costs of a department or hospital and the theoretical income based on the new financing system are a first step to gain more insight into the efficiency of hospitals.

Because this model is strongly data based, attention has to be given to the data quality. Therefore some comparisons especially about the number of medical procedures and additional diagnosis have been performed. Furthermore, checks for the plausibility of diagnosis and medical procedures have been carried out. These analyses show that at the moment big differences in data quality between hospitals exist. But from the experience of the last years we can observe a continuous improvement of data quality. But a lot of work has to be done to reach a global high standard in documentation to use this data for the new financing system.

A Steering System as Part of the Hospital Financing System

The new hospital financing system should work as a steering mechanism and should have some regulatory effects. It should also be a first step in a data-based planning of future investments in hospitals. But it should not have negative effects on the quality of care and on the high level of medical assistance in Austria. Since the beginning of the development of the model there have been discussions on whether the LDF-points are sufficient or if additional variables have to be considered. In view of the large variety of structures within hospitals and special functions of some, it was decided that additional factors should be taken into account. That means that special functions of a hospital – such as supply functions, schools for nurses, academic training, etc. – should be considered. Also in regions not easy to reach the rate of capacity utilization can be low and therefore a special support is necessary. In general this steering mechanism should be related to the quality of care, it should be easy to understand and the reproducibility should be high. It should include incentives to increase structural quality and as a consequence the quality of care.

To develop a transparent model that also includes the steering mechanism, data for the elements of the steering system have to be available. But at the moment good data bases for structural quality, training facilities or hotel quality are not available. The most valid data we have are about the different professional groups. Regression analysis has shown that the number of medical doctors and nurses per bed are good explanatory variables for the costs as single variables, especially if they are combined with the LDF-points. This is not surprising since personal costs are the main part (60%) of the whole running costs. The goodness of fit of predicted costs to the actual costs improved significantly if indicators about the staff per bed is included. But these factors can also be criticized because there is no clear relationship to the performance or service in a hospital. On the other side, it is well known that hospitals with special high-tech service functions need more staff.

We decided to use a non-linear logarithmic weight function of the number of medical doctors and nurses per bed and formed a weighted sum of medical doctors and nurses. To study the influence of this personal factor to predict the costs a certain amount of the whole costs (15%, 25%, 50%) was distributed according to this factor and the remainder according to LDF- points. This model increased the reproducibility of the actual costs. But the importance of this factor and the amount which is distributed by this factor has to be discussed from a quality viewpoint. Another approach, which is under consideration now, is to give extra charge for hospitals with a staff above a certain limit and to deduct a certain amount for hospitals below a certain limit.

For all these calculations the value of the LDF-point is the same for all hospitals. An analysis of the goodness of fit of the model suggests that the type and function of a hospital, like university hospitals on the one side and hospitals for basic care on the other side, should be considered and the value of a LDF-point is weighted by this factor. Because there is no explicit information about the severity of illness in the data, this additional adjustment can substitute this information under the premises, that in some type of hospitals the severity of illness of patients is higher than in other ones.

AIMS AND POSSIBLE EFFECTS OF THE MODEL

By introducing this new financing system some regulatory effects are expected. The aim of this regulation is that a hospital should be economically and technically efficient, produces a maximum output for the input it uses. A fixed price per admission system is considered as a more or less direct method to prevent excess capacity. Hospitals have to take care to bring a certain service at a given reimbursement rate. If there is excess capacity, facilities have to be closed, because the fixed reimbursement rate would not cover the costs (Feldstein, 1993).

The aims of the new financing system are to:

- stop the cost increase
- increase the efficiency within a hospital
- pay in an adequate manner for diagnosis and therapy and not for the length of stay
- guarantee free admission for all Austrian people to hospitals and especially also to expensive procedures
- improve the quality of care
- guarantee medical progress
- decrease the length of stay (as a consequence of the new financing system).

It is expected that the model will have some regulatory effects on the whole health market. At the moment we do not have any practical experience with this new financing system but it seems necessary to discuss possible positive and negative effects in advance (Andersen Consulting, 1993; Kahn et al., 1988).

From the patients' perspective, both advantages and disadvantages have to be expected. The advantages are:

- shortening of the length of stay
- higher efficiency for diagnostic procedures, especially reduction of multiple examinations
- development of guidelines and standards
- better information.

The disadvantages are:

- early discharge ("the quicker the sicker")
- neglecting or avoiding of necessary tests and diagnostic procedures

- increase of re-admissions
- additional procedures to shift a patient into a better charged group ("DRG-creeping")
- patient shifting and selection of profitable patients ("cream-skimming").

Both in the actual and the proposed system there are no incentives concerning cost awareness of patients.

From the perspective of the hospitals' employees the advantages are:

- increase of efficiency (by system analysis to optimize the patient path)
- improvement of the documentation and communication system
- development of guidelines and standards
- awareness of costs.

The disadvantage for this group is:

- cost control (selection of cheaper and not optimal products).

From the management of the hospital perspective the following advantages can be expected:

- prospective calculation of the costs
- analysis of efficiency
- reduction of the number of beds (or more patients)
- technology assessment
- open to innovations with reduction of the length of stay
- more specialization (not everything can be done in any hospital)
- co-operation within a smaller region.

The disadvantages are:

- competition within a smaller region
- cost awareness.

The reputation of a hospital will become very important and the communication and co-operation with general practitioners will improve. The rate of assignment will become an important factor for the rate of utilization.

Lastly, the following advantages can be expected from the health system's perspective.

- reduction of the increase of costs
- comparison of services (small area variation)

- data based planning of the health system and of expensive equipment
- development of guidelines and standards
- more ambulatory, less inpatient care
- more information about costs and outcome.

The disadvantages are:

- limitation of services
- "cream-skimming" of competing suppliers
- additional nursing homes for post hospital care.

As the new financing system depends very much on the data quality, hospital information and communication systems and documentation will play a very important part (Andersen Consulting, 1993). The introduction of hospital information, communication and expert system is one of the most promising factors in cost reduction of the future. It is expected that data quality will increase substantially after the introduction of the new hospital financing system. But we also have to be aware of possible manipulations, such as the DRG creeping. Furthermore, it will become necessary to introduce internal quality management and to develop procedures for external quality control. An efficient external quality assurance system is necessary to avoid low quality of care and to guarantee well-defined quality standards.

DISCUSSION

In the last years we have prepared a new hospital financing model in Austria. This model is based mainly on the diagnosis, special procedures and some characteristic values of the patient; but it is independent from the length of stay. To consider the structural and hotel quality as well as other important factors of health service in certain regions an additional steering system was developed. Homogenous groups of patients identified from a large learning sample of patients of Austrian hospitals using modified regression tree algorithm reflect to a large extent the characteristics of the Austrian medical culture and standards.

For each acute hospital in Austria the actual costs have been compared with the costs calculated with this new model. There is a very good reproducibility globally, but for some hospitals the differences are big and explanations have to be found. An additional steering system is necessary to handle structural problems. Although we do not have any practical experience with this model, many discussions have shown that it is a very valuable instrument for hospital managers and health service analysts to compare the service spectrum of hospitals, to study efficiency within hospitals, and to perform data-based planning.

The reactions of the hospitals depend of course on the structure of the owner, the ability of the management and the available infrastructure. Today many public financed hospitals are more administered as managed by well devised strategic planing. Hospitals, and especially inpatient care are the most expensive factor in the health system and a change in the hospital financing system will affect all other institutions. Therefore in the near future similar financing models have to be developed for ambulatory care. The global aim is health system efficiency and therefore the entire system has to be considered. This includes also that hospital care, relative to other forms of care, is in an appropriate relationship. To minimize costs means also to substitute hospital care by less costly institutions (Feldstein, 1993). Holding down hospital costs may also lead to an earlier discharge of patients and providing fewer services and can therefore lead to a shift of costs (Kahn et al., 1988).

In a next step it is necessary to establish this new financing system not only for a part of the operating costs but also for the investment costs. One way is to go into the direction of monistic financing by creating more incentives for organizational changes and for more competition.

ACKNOWLEDGEMENTS

I want to thank Mag. G. Embacher, Dipl.-Ing. H. Gaugg (Ministry of Health) and Univ.-Prof. Dr. D. Klingler (AKH Linz) for many stimulating discussions and Dipl.-Ing. B. Pesec for the development of the computer programs.

The work was supported by and partly performed on behalf of the Austrian Ministry of Health, Section II.

REFERENCES

Andersen Consulting (1993). *The Future of European Health Care*. London: Andersen Cons. and Burston-Marsteller.

Feldstein, P.J. (1993). *Health Care Economics*. Albany: Delmar Publishers.

Kahn, L.K., Draper, D., Keller, E.B., Rogers, W.H., Rubenstein, L.V., Kosecoff, J., Sherwood, M.J., Reinisch, E.J., Carney, M.F., Kamberg, C.J., Bentow, S.S., Wells, K.B., Allen, H., Reboussin, D., Roth, C.P., Chew, C. and Brook, R.H. (1988). *The Effects of the DRG-Based Prospective Payment System on Quality of Care for Hospitalized Medicare Patients*. Final Report. Santa Monica: RAND.

Österreichisches Statistisches Zentralamt (1992). *Statistisches Jahrbuch für die Republik Österreich*. Wien.

Pfeiffer, K.P., et al. (1991). Identification of groups of patients consuming equal resources for a hospital financing system using regression trees. *Medical Informatics Europe, Vienna*. Berlin: Springer, pp. 845–9.

Pfeiffer, K.P., et al. (1994). *Stability of Regression Trees. Compstat*. Vienna: Physica-Verlag, pp. 166–71.

Chapter 14

Fixed Budgets for Fundholding General Practitioners in the UK

Howard Glennerster

SOME BACKGROUND

The United Kingdom's health reforms in 1991 were not designed to introduce an overall budget limit for the National Health Service (NHS) or even set local fixed budgets. These kinds of budget limits had been in place for many years in the UK as Clive Smee (Chapter 11) has shown. Sometimes British politicians have been more generous in setting those ceilings than at other times – they tend to be more lax before general elections, for example – but Her Majesty's Treasury has always maintained tight overall control of health spending in the UK.

The pre 1991 system did not, however, contain any incentives for efficient behaviour by hospitals or GPs within those budget limits, indeed the reverse. Treating more patients did not in any very direct way attract to a hospital or gain its senior doctors any additional resources. To go slow on the treatment of your non-emergency patients would encourage them to attend your private clinic. There was no financial incentive for GPs to be careful in the prescribing of expensive drugs. The basic purpose of the reforms was to redesign these incentive structures, to use them to get more out of the existing service and hence to sustain the very tight ceilings that were proving politically embarrassing to hold.

Local hospitals were no longer to receive a fairly automatic incremental budget allocation. Instead, they would have to compete for funds given out on a contract basis by district health authorities. The more work they did of the precise kind the district wanted, the more money they would get. The hospitals would become free-standing agents dependent on attracting custom.

In practice, this reform was less radical than it seemed, since many districts had to rely on the one large local hospital to do most of their work. There was one part of the reforms, however, that took the whole competitive and devolved budget logic a stage further. Money that used to go direct to hospitals was given instead to family doctors

Fixing Health Budgets: Experience from Europe and North America.
Edited by F. W. Schwartz, H. Glennerster and R. B. Saltman. Technical editor R. Busse.

who could use it to buy hospital care for their own patients.

The UK's experiment with these fundholding GPs, as they are called, is one of the most radical attempts to change the structure of a publicly funded national health care system anywhere in the world. It is this part of the reforms that is the subject of this paper. Fundholding was radical for four reasons. First, because it switched the funding of health care from a top-down model of bureaucratic allocation to a bottom-up system which put general practitioners in the financial driving seat.

Second, since patients can choose their own GP, and measures were taken to increase this freedom, patients could choose who purchased hospital care on their behalf. Unlike the main model of reform in the UK, where districts have purchased on behalf of their whole populations, this introduced the possibility of competition between purchasers as well as providers.

Third, it amalgamated, in the hands of a single budget holder, spending on primary health care, hospital care and community health services, all previously in the hands of separate bureaucracies. It thus opened the way for some questioning of the boundary lines between these services and for moves that could improve the allocative efficiency of service delivery.

Lastly, the scheme introduced a fixed budget for the pharmaceuticals that a GP could dispense in any one year to patients on his or her list. With this budget constraint went the incentive to cut pharmaceutical spending since any spending below the cash limit could be spent on other forms of patient care by the practice. GPs could, for the first time, trade off prescribing drugs against other forms of treatment. Again, this opened up the possibility of better allocative decisions given appropriate information.

Here were four reforms for the price of one. But they only applied to those patients whose GPs were willing and able to join the scheme.

It is not surprising, then, that the fundholding scheme has become the most controversial part of the reforms to the NHS and the most researched element within it – though that is not saying much! (For the only comprehensive review of research on the NHS reforms see Robinson and Le Grand, 1994; see for comment Ham, 1993a, b; for a review of the research on fundholding see: Dixon and Glennerster, 1995.)

The most resented and controversial aspects of fundholding relate to the fact that the scheme only applies to those GPs who chose to join the scheme. Any advantages it might have are therefore not generally available. The second main criticism concerns the disputed claim that these GPs are getting more money per patient than are the districts who purchase on behalf of non-fundholding GPs. I do not propose to discuss either point in this chapter. The solutions to both problems are obvious. If it proves to be intrinsically better, the scheme should be extended to the whole population, not abolished. If it is more generously financed, then government should make sure all GPs and services are financed on a common and fair basis.

The real debate ought to be about the intrinsic merits and dangers of this form of funding. That is what this chapter seeks to address.

There are four basic questions to ask about the scheme:

1. What is the scheme?

2. What is it trying to achieve?

3. What evidence is there about its success in terms of these objectives?

4. What are the inherent dangers, if any?

THE SCHEME

The Original 1991 Scheme

Under the original fundholding scheme (DoH, 1989) introduced in 1991, family doctor practices with more than 9000 patients could apply for fundholding status. Regions drew up a list of criteria to screen applicants in order to ensure that the practices were managerially and technically capable of handling the scheme and were fully committed to it.

These practices received a budget allocation that could be spent only on a defined set of purposes. It could not, in theory, be used to increase the doctors' income or to benefit the practice generally. This budget was not paid as cash to the practices but held by the Family Health Service Authorities and released to pay hospitals when authorized by a practice. The drugs prescribed by doctors were not paid directly out of any cash the doctors held but were set against the cash limit fixed for the practice. Any savings under that limit could, however, be turned into an enhanced budget allocation for other elements in the budget. The intention was to put a cordon sanitaire around the fundholding budget, to keep it separate from the wider income earning activities of the partnership. The budget allocated to the practices to buy hospital services was deducted from the allocations the district health authority would otherwise have received. From April 1991 to March 1993 the budgets covered the following expenditure:

- *Hospital inpatient care for a restricted range of operations.* These covered: ophthalmology; ear nose and throat; thoracic surgery; operations on the cardiovascular system; general surgery; gynaecology; orthopaedics. Emergency referrals in the above categories were excluded.

- *All outpatient visits.* These were defined as "a patient attending a hospital or clinic other than as an inpatient" or a home visit arranged by the GP. Excluded were first appointments made where treatment followed a period as an inpatient but not in the fundholding list of treatments. Chemotherapy, radiotherapy (for example, for cancer), renal dialysis and genito-urinary clinics were excluded. As for inpatient care the budget was set on the basis of an estimate of the practices' past use of these services.
- *Diagnostic tests done on an outpatient basis.* These included blood tests, urine tests and X-rays. Breast screening under the call and recall scheme was excluded. Budgets were set on the basis of evidence collected in the previous year about the level of activity practices demanded from hospitals. The number of different types of test requested in previous years were multiplied by their expected price.
- *Pharmaceuticals prescribed by the practice.* The drug budget was calculated in exactly the same way as the indicative budget within which non-fundholding practices were expected to keep.
- *Practice staff.* Again fundholders were paid the same as other GPs for their practice staff but instead of having to get approval for each staff member, and new approval each time a receptionist left, for example, fundholders are simply given a sum of money to cover staff costs. This is added to their total budget. It gives them flexibility to save money on staff and add savings to other elements in their budget or to make savings elsewhere and employ more staff.

Though calculated separately all these budget elements were pooled. A saving on any one of them could be used to spend more under another heading. Initially the budgets were calculated in such a way that each practice was able to refer as many patients to hospital as it had in the past and to prescribe as many drugs. This base line budget was then rolled forward with inflation subject to individual practice by practice negotiations. More recently, in line with long-term intentions, the budgets are moving to a capitation formula basis reflecting the size and sex and age structure of the practice's population. In addition the practice is paid a management allowance to cover the cost of additional management and computing analysis needed.

Careful track is kept of budget authorizations and actual spending out turns by the local and regional regulatory bodies. If potential over spending is a danger the practices can be warned. If overspending does happen in one year, the next year's budget can be adjusted to recover the sum or adapt the budget if special factors were at work. However, if there is over spending in the next year too, the whole budget will be taken away.

The purchase of hospital services by these GPs covered only about 15% of hospital activity in cash terms. In other words, if all GPs in an area had opted for GP fundholding status they would have controlled only 15% of the budget of the local hospital services.

1993 Extension

From April 1993 the scheme was extended to include practices with only 7000 patients. Two practices with 3500 patients each could combine for fundholding purposes. The scope of the fund was also extended to include:

- *Community services.* This includes: district nursing; health visiting; chiropody; dietetics; all community and outpatient mental health services; mental health counselling; health services for people with learning disabilities; all referrals made by health visitors, district nurses and community mental handicap nurses. Terminal care and midwifery were excluded. The health visiting and district nursing budget covered services provided for patients of the practice as well as referrals to other services made by these staff and the costs of equipment.

In 1994/5 the allocations to fundholders were made up in the following way (National Audit Office, 1994).

Hospital and community services purchases	56%
Drugs budget	40%
Staff	4%

Budgets were to be set by regions assuming the existing levels of service plus any planned changes such as a substantial increase in the list size. Each practice would be funded for the existing level and grade of staff allocated to it. Practices would not be charged for the public health duties health visitors performed. When calculating the prices to be charged to practices community units must include the costs of management and specialist support and training. The Department of Health (DoH) ruled that practices could only purchase from established NHS community units. GPs would not be allowed to employ community staff directly or through the private sector.

1996 Extension

In October 1994 the Secretary of State announced further extensions to the scheme. Practices with 5000 or more patients will be able to join the scheme from 1996. Practices with only 3000 patients will be able to join a scaled-down scheme covering only drugs, community services, staff and diagnostic testing. There are to be experiments in about 60 sites of total fundholding which means that groups of practices could combine together to purchase all hospital and community care, making general practitioners the purchasers of health care for all their patients' needs – essentially a full HMO model (NHS Executive Letter 79/94). At the time of writing (summer 1995) the experiment has proved even more popular with GPs than originally thought. There is to be a national evaluation of the experiment lead by the Kings' Fund.

The Scheme's Growth

In the first year of the scheme 7% of the population were covered by GPs who joined the scheme. In the second year the scope of the scheme nearly doubled, covering 13% of the population. By 1995 40% of people in England were registered with fund-holding GPs. The population covered differed from region to region. Within regions, the spread of fundholding was even more varied. In one county we studied nearly two-thirds of the population were covered in 1994. In other areas a fundholder was rare.

The scheme has therefore developed rapidly despite its radical nature. This has not dimmed political controversy. The Labour Party is committed to abolishing the scheme, essentially because of the inherent inequity of having two funding mechanisms running side by side, though it has proposed introducing ways of involving GPs more directly in the purchasing or commissioning function (Labour Party, 1995). The long-term future of the scheme is therefore in doubt. The present government sees fundholding as the way forward. The main Opposition parties see it as anathema.

THE SCHEME'S PURPOSE

It is always difficult to know exactly what the origins and purpose of any political reform are, though, the origins of this reform can clearly be traced back to ideas floated by Alan Maynard in the early 1980s (Maynard, 1986). What follows is an attempt to put some kind of economic logic to the normal confusion of motives and intentions that go to any policy change. (For a fuller account see Glennerster et al., 1994.)

The Logic Behind Fundholding

Large-scale monopsonist purchasers of health care services have no particular incentives to contract on behalf of their populations or reflect their preferences. They and their employees are not at risk if they do a bad job. They have nothing to gain if they do a good job. Employees do, however, have an incentive to minimize the effort involved in contracting. This could mean setting sweetheart contracts with the local large provider (Propper, 1993). (In the UK district health authorities are not even elected so they did not face electoral incentives either.) If the competitive logic carried any weight, the argument goes, it should apply to the purchaser side of the market as well as to the provider side. If patients were given increased power to choose between GPs and they were given the power to purchase services, patients would then be able to choose between alternative purchasers. Since GPs derive their income from the number of patients they can attract as ordinary GPs they would have an interest in being good purchasers. The better deal they could get for their patients the more attractive they would be as GPs and the higher their potential income.

GPs have other reasons for wanting prompt and good service from hospitals and community services for their patients. The longer the patient has to wait for treatment the more time he or she spends in the GP's consulting room taking the GP's time and causing frustration and stress. The GP gets more professional satisfaction, wastes less time and suffers less stress if a patient is seen promptly. These are strong motives for getting good speedy contracts. They are also good reasons for wanting the patient to be satisfactorily dealt with. A patient who returns either unsuccessfully treated or with post-operative complications causes the GP trouble and time. Nothing comparable can be gained by a district manager. Professional satisfaction with a good epidemiological study well done or a sound contract negotiated are real but not as important in the end as the minimization of time and frustration and stress. GPs have stronger motives to be good contractors on behalf of their patients. This may be called the public choice-based case for fundholding.

Critics of fundholding have rightly pointed out that a GP has no particular incentive to make contracts on behalf of those in the community who may be in need but do not come forward and seek care – pregnant mothers in some communities, for example – or a drug addict or a homeless person. But neither does a non-fundholder or a district purchaser. If we want to encourage more such preventive action FHSAs have to give incentives to make it happen, fundholding or no fundholding.

Not only does a GP have the motivation to seek better contracts for the range of services included in the scheme but he or she will be more likely to possess better detailed information about the quality of work done by individual hospital specialists. A GP will have experience drawn from regular contact, discharge letters and other communications about individual cases. This information may be anecdotal rather than statistical but, then, so is most purchasing in the marketplace. Districts or the new larger health commissions have the staff qualified to collect statistics on need and on hospital performance. They can then make it available to those who are doing the purchasing. A GP is, thus, always in potentially a superior information position to a district manager. The GP has, or can have, both patient-specific data, hard and soft, and statistical or epidemiological data analysable on a larger comparative basis, so long as the district or other public health analysts can supply it. In the new total fundholding schemes, that expertise can be directly employed by the total fundholding group.

In short, GPs have, potentially, both superior information and the motivation to use it in contracting compared with districts or larger purchasing or commissioning agencies. The economics of bureaucracies and contract theory suggest that the GP should be a better purchaser, at least for those conditions he or she deals with relatively frequently and where there is a genuine choice of provider. Nevertheless, GPs also have disadvantages as purchasers.

Disadvantages of Fundholding

First, general practitioners operate in very small groups compared to the risks and size of unpredictable budgets.

Second, the ordinary GP has very little experience of treating and therefore contracting for a wide range of rarer conditions. GPs are not in a good position to see the whole needs of their areas as opposed to the restricted group, who might be their patients or those patients who came to see them on a regular basis.

Recognizing this the original reformers confined the budgets to non-emergency and relatively cheap treatments frequently met by GPs. They also introduced a stop loss provision so that if a patient cost more than £5000 (later £6000) a year the cost would be met by the district. A manager could be employed to handle the fund. This logic explains the relatively restricted nature of the scheme when it was first introduced and the size restrictions on the practices permitted to join.

Third, the American experience with HMOs suggested that competition between purchasers financed on a prepayment system would lead to GPs cream skimming the healthy patients, excluding the potentially costly ones. It also suggested that setting the capitation formula would be very difficult, especially if it were to seek to address this problem (Epstein and Cummella, 1988; Howland et al., 1987; Lubitz, Beebe and Riley, 1985; Newhouse, 1986; Newhouse et al., 1989).

SUCCESSES AND DANGERS: METHODOLOGICAL ISSUES

Economic theory and previous experience of competitive markets in health care thus predicted both potential costs and benefits from bottom-up purchasing. So it has proved. It has proved difficult to reach any consensus on the scale of the costs or the benefits and hence to draw up a balance sheet. I was fortunate to be awarded a grant from the Kings' Fund in London in 1989 to study the introduction of the scheme and was able to follow a group of ten fundholding practices through their first four years and a group of 16 practices who joined the scheme in the third wave. The detailed results of this study have been reported (Glennerster, Matsaganis and Owens, 1992; Glennerster et al., 1994; Matsaganis and Glennerster, 1994; Robinson and Le Grand, 1994). It was a study primarily designed to test the administrative feasibility of the scheme not its health outcomes.

Few studies in this field have attempted a randomized or control group comparison except of certain narrow features of the practices' experience. This is not surprising. Not only are there the normal problems with trying to apply such techniques to the messy and changing world of policy, but there are intrinsic difficulties with such an approach in this case. If the government had been prepared to develop the policy in some parts of the country and not others the story might have been different. But governments that think they have a good idea are not ready to wait. Fundholding therefore proceeded alongside district-based purchasing throughout the country. The very theory behind fundholding predicts that the enhanced competitive environment will have an impact both on providers and on other purchasers – districts. They will have an incentive to improve their purchasing if fundholders are seen to do a better job. The improvements gained by fundholders in their contracts will be claimed by districts in their negotiations and vice versa.

Our conclusion was that the field study approach, which went out of fashion in economics in the 1940s, was the method that was most capable of throwing light on the market situation and the relative performance of the actors. Since the case for the two systems rested in large part on the efficiency of the two kinds of purchasing, it would be worth comparing the process of purchasing or contracting in the two systems in some detail and looking at the outcomes. What did the two kinds of purchaser want to achieve? Was there any evidence that GPs sought different things? Were they more successful in getting their way? Did they have more leverage over hospitals? Did the hospitals respond at all differently? Thus, a central part of the study compared the process and outcomes of contracting as undertaken in the same areas for the same range of services by districts and by GP fundholders.

Like all experiments, such research was liable to the Hawthorne effect. In fact, though, the results of the contracting were so important to the practices' everyday lives and finances that our presence was likely to be very marginal. Worse was the effect of the rigorous selection of practices in the first wave. They were all especially good practices. Were we only seeing the effect of their quality? We countered this to some extent by repeating the survey two years later with third wave practices. In fact, we found the effects we observed more powerful in the third than the first wave.

Different again was the problem that our practices might be getting good deals just because they were so marginal to the whole process in the first years. Perhaps hospitals were offering them especially good marginal cost deals. If this were true it could be tested by comparing the negotiating process in areas where few GPs were fundholders with that in areas where most GPs were part of the scheme. There they could not be gainers at others' expense. An alternative hypothesis was that the more fundholders there were in any given area the more leverage they would exert on providers just because of their market power. As we shall see there was evidence of some marginal cost deals in the first year but the most powerful effect on service quality was achieved by GPs in the area where they controlled virtually all the budget. They were instrumental in getting two consultants removed about whom there had been repeated complaints over many years, for example. (A highly unusual event, it has to be said, in the NHS.) Evidence from close observation of the contracting process was reinforced by the views of GPs who were opting in increasing numbers for the scheme. When asked why, most gave as their main reason their view that the fundholders had been better at getting what they had long wanted from the local hospital service.

Another kind of evidence came from the fact that districts began to involve non-fundholding GPs in their contracting in much more direct ways. These schemes were very varied and ranged from asking GPs' views more assiduously, to actually giving them a major role in contracting or even a shadow budget. These schemes go by the name of locality purchasing or GP lead commissioning (Ham, 1993c). If copying is the sincerest form of flattery, this was some evidence of the appeal of the bottom-up approach. We are in the midst of a study that compares fundholding with locality purchasing, GP lead commissioning and this multitude of variants.

Other claims made by proponents and opponents were easier to measure – that the budgets would be volatile and lead practices into bankruptcy. Here we were able to do a straightforward budget monitoring exercise and an analysis of budget spends against target. It did not prove a significant problem. Because the fund covered only non-emergency and relatively cheap hospital procedures GPs were able to speed up or slow down the size of the queue to keep within budget. Most came in within 1.5% of their target by year two.

The most troublesome prediction, drawn from the economic literature, was that practices in competition for patients would compete to select out the least costly patients. The incentives inherent in the UK scheme are not the same as for HMOs in the USA, however. In the first place, in the early years of the scheme, the practices were financed on a historic cost not a flat premium basis and the stop loss of £5000 minimized the impact of expensive patients. In the second place there was no direct financial risk for doctors suffering from the consequences of having a high-cost patient. The fundholding budget is completely separate from the GPs' remuneration and can only be spent on patient services. The losers in a practice that found it had high-cost patients would be other patients. However, if the patients of a practice with more costly patients began to get worse service and left the practice this would affect the GPs' income. Thus, the incentives to cream skim are limited but still real. Given the chance to avoid budgetary problems a GP is likely to want to do so.

We were able to test whether there were incentives for cream skimming and how big they were. We did this by taking a random sample of patients from one practice's list and tracing their use of service and the costs. We were also able to abstract from the patient records predisposing factors that might be linked to high costs. If we could do it, so could a GP from his or her own computer records.

We were able to show that certain kinds of patient were predictably more expensive in fundholding terms, i.e. in the demands they would make on services coming under the fundholding scheme. Age and sex were obvious categories but so were long-standing conditions and previous hospitalization. Statistically significant results for a series of simple predisposing measures emerged using methods developed in the USA.

(This part of the study is reported in Matsaganis and Glennerster, 1994.) There were, in short, both predicted efficiency gains and equity losses to be derived from this radical new scheme.

SUCCESSES AND DANGERS: RESULTS AND COMMENTARY

The Efficiency Criterion

Hospital Efficiency

Even if every GP were a fundholder, less than a fifth of hospital expenditure would be purchased by them. Thus the impact of the scheme on hospitals, for good or ill, was likely to be limited. However, GPs' purchasing power in the fundholding scheme was concentrated on those parts of the hospital service that have performed least well over the years – elective surgery, waiting times, pathology and outpatient care.

Some of the earliest and most dramatic results were obtained in pathology. The internal incentives for monopoly hospitals to provide good service for GPs were minimal. The service was often very poor, with doctors waiting over a week for test results. In contrast, once opened up, the potential market was highly contested with both private firms and hospitals over a very wide distance able to enter the market. Fundholders in our sample, by threatening to move from existing providers and changing them where necessary, were able to get their tests in a day rather than a week.

After comparing the contracting process and outcomes secured by fundholders with those obtained by districts we concluded that the evidence did support the view that in the first three years of the scheme fundholders had been more successful in affecting hospital behaviour than the districts in the same areas. The GP's contracts were more reflective of individual practice needs, were more innovative and reflected more micro patient concerns. The results of this more detailed analysis of the contracting process are contained in our study (Glennerster et al., 1994). Since the early days districts have come to be more aggressive and to involve GPs more in their contracting process but they have been driven significantly by a desire and a need to do as well for their GPs as fundholding GPs have been able to do for themselves.

Allocative Efficiency between Sectors

The second hypothesis was that combining primary and secondary care budgets would enable services to shift to the most convenient site. The removal of budget boundary walls between hospital, practice and community services has begun to produce more flexibility and a growth in practice-based work. There is no one right balance but discussion about the appropriate balance and the capacity to do something to change it, is evident. Fundholding has been a catalyst. Devolved budget responsibility does indeed seem to release innovation.

Practice Efficiency

One of the by-products of the additional managerial input needed to run the fund was improved practice efficiency. The rest of the practices' activities came to be better managed too.

Economy on Drugs

Part of the scheme which is often overlooked is the inclusion in it of an element that covers the cost of drugs prescribed by the GP. If the practice saves money on this budget it can be used in the rest of the fund. This gives a direct incentive to GPs to review carefully their prescribing habits in a way that government and commentators have been urging. The evidence from our regular interviews was that our sample GPs did devote serious attention to containing their drug budgets. They did so more than non-fundholders who did not face such direct incentives. We did not possess randomized control data to compare our practices' resulting drug spending with non-fundholders. However, two studies with matched samples of fundholders and non-fundholders suggested that fundholding gaps have not reduced the costs of their prescribing but have increased the costs of their prescribing more slowly than non-fundholders (Bradlow and Coulter, 1993; Maxwell et al., 1993). One other study suggested that practices had done this by using more generic equivalents to brand name drugs and to have been more cautious about taking up expensive new treatments (Burr, Walker and Stent, 1992). A more recent study of eight first-wave practices suggests that the effect may be fading (Stewart-Brown et al., 1995). National figures show the gains being sustained, however (DoH, 1995).

Administrative Costs

The costs of practice-based contracting are clearly higher than the district-based contracting. The total cost of fundholding per practice, including the costs to hospitals and FSHAs and Regions as well as to the practice, amounts to about £80 000 a year (Davies, 1995). This is roughly 5% of an average fundholding budget. The additional costs are to some extent offset by the clinical advantages that flow from the improved individualized patient information and monitoring of hospital progress that can be undertaken by the practice as a result. But there is no doubt that GP-based contracting costs more. The question is whether the outcomes are that much better.

One small cost-benefit study done in one of our practices suggested it might be. The practice had negotiated outpatient work in its surgery. Calculation showed that this one clinic on its own had saved patient travel time worth the additional cost of managing the fund without any other benefits.

The Scale of Referrals

One original intention that lay behind the thinking about GP fundholding was to make practices think more about the reasons for referring patients to hospital by facing the GP with the costs of each referral. If every GP were given similar budgets it might in the end reduce the wide and unexplained variability in GPs' referral patterns. In fact, the way the GPs' budgets have been set has not helped to achieve this goal. Since budgets were initially set on the basis of existing referral rates GPs have had no reason to revise their habits in this respect. Coulter and Bradlow (1993) have shown that this was so in Oxfordshire, in one of the most careful studies in the literature. This strengthens the efficiency case for moving to a capitation based funding allocation.

Comprehensive Planning

Even if there are micro efficiency gains to be reaped from having competitive purchasers, critics argue, the result is that districts lose their capacity to plan services for their populations. That requires one purchaser for the whole area. GP fundholders have fragmented any capacity to do this. They respond to patients' demands not the population's needs (Ham, 1992). They are less concerned with meeting the goals of The Health of the Nation (DoH, 1992). There are both a series of answers to these claims and some truth in them.

First, we have to remind ourselves yet again that, even with the inclusion of community services and 100% coverage, fundholders potentially control no more than a quarter of the combined hospital and community services' budget. In practice they have, in the past two years, only controlled 2% and now about 7% of that budget, and control over the community services element has been minimal due to government restrictions on the way it is administered.

Second, the concept of planning that districts tend to use in this argument is rooted in epidemiology and the past. The task of a district planner is seen to be examining trends in disease and populations at risk. Consumer preferences and patient demand are rude words or, at least, alien concepts. If they are to be used at all they involve patient questionnaires. Private service organizations, however, have a quite different notion of planning. So, increasingly, do the new "enabling" and contracting Social Service Departments. An organization such as a supermarket chain uses trends in consumer preferences to tell them what consumers want. As we saw in several of our areas, districts were using the market choices of their GPs as valuable behavioural information in planning future services.

My own view is that if purchasing is to be GP-led to a much larger extent, as both the present government and the Labour Party in their different ways are saying, then the epidemiology input has to be effective at the GP fundholding or commissioning level. Indeed it should be more effective there. Where else can Health of the Nation goals be delivered except by primary care (DoH, 1992)?

We found that fundholding practices were more aware than they used to be of health service wide issues. They were more aware of financial issues and participated more in discussions about local priorities. This somewhat unexpected finding is consistent with Howie's (1992, 1993, 1995) findings in Scotland.

Equity and Future Development

The equity side of the balance sheet is more worrying. Some work has suggested that fundholders received higher budgets than the implied budgets of non-fundholders in one region (Dixon et al., 1994). This is often claimed by opponents of the scheme. Another region undertook a survey that showed the opposite, as did early national figures (Glennerster et al., 1994).

A second worry concerns the old problem of cream skimming. Where doctors or health providers are paid a flat sum for each patient they look after they will tend to cream off the most potentially expensive ones (Luft, 1986; Luft and Miller, 1988; Newhouse, 1982; Robinson et al., 1991; Welch, 1985). Our study suggested that there was real scope for cream skimming (Matsaganis and Glennerster, 1994). Two-thirds of the fund spending went on 5% of the patients. As yet there is no hard evidence of cream skimming actually happening. As formula funding becomes the norm, it could do if the formula does not fully reward practices for taking predictably more expensive patients. At the moment age and sex are in the guideline formula. However, there is also scope for countering these incentives further with a capitation formula that meets the predicted higher costs likely for patients suffering from a few key conditions.

Colleagues at York have used a different methodology based on small area statistics and argue that patient-based methods of formula funding are not feasible (Sheldon et al., 1994). I still disagree. We have demonstrated that even if age, sex and social deprivation are taken into account individuals in certain obvious categories, like diabetes, for example, are likely to cost the practice budget a great deal more than others.

A practice could receive a supplementary budget if it informed the authorities that it had more than a set percentage of patients on its list with defined medical conditions or histories. My experience of GPs is that they would be capable of generating this information if it brought financial reward. The information would then have to be sample checked like the US Revenue Service do and draconian penalties imposed for cheating.

If we are to go down the road to total fundholding, or indeed locality commissioning, and there is to be competition between practices in an area then the formula becomes critical. Not only does the old problem of budget volatility rear its head again but so would the threat of cream skimming on a much larger scale. With a limited budget, and a low stop loss insurance on a patient costing too much, these problems were containable. That would no longer be the case with a budget that covered all medical care. The experimental schemes chosen seem to be based on large groupings of GPs. But in solving one problem the large groups lose the competitive characteristics that may have given the scheme its particular virtues.

There is clearly much more work to be done here. If we are to move further down the route to GP-based purchasing – either under the present system or a Labour government version – these funding issues become critical.

REFERENCES

Bradlow, J. and Coulter, A. (1993). Effect of fundholding and indicative prescribing schemes on general practitioners' prescribing costs. *British Medical Journal*, **307**, 1186–9.

Burr, A.J., Walker, R. and Stent, S.J. (1992). Impact of fundholding on general practice prescribing patterns. *Pharmaceutical Journal*, **24**, October.

Coulter, A. and Bradlow, J. (1993). Effect of NHS reforms on general practitioners' referral patterns. *British Medical Journal*, **306**, 433–7.

Davies, J.A. (1995). Survey of the administrative costs of fundholding. *Fundholding*, January, pp. 22–4.

Department of Health (1989). *Working for Patients*. London: HMSO.

Department of Health (1992). *The Health of the Nation*. London: HMSO.

Department of Health (1995). Figures supplied by the Economic Adviser's Office.

Dixon, J. and Glennerster, H. (1995). What we know and what we ought to know about fundholding. *British Medical Journal*, September.

Dixon, J. et al. (1994). Distribution of funds between fundholding and non fundholding practices. *British Medical Journal*, **309**, 30–4.

Epstein, A.M. and Cumella, E.J. (1988). Capitation payment: using predictors of medical utilization to adjust rates. *Health Care Financing Review*, **10**, 1, 51–69.

Glennerster, H., Matsaganis, M. and Owens, P.A. (1992). *Foothold for Fundholding*. London: Kings' Fund Institute.

Glennerster, H., Matsaganis, M., Owens, P. and Hancock, S. (1994). *Implementing Fundholding Wild Card or Winning Hand?* Milton Keynes: Open University Press.

Ham, C. (1992). *Learning from Experience: The Next Steps in Implementing the NHS Reforms*. Birmingham: Health Services Management Centre, University of Birmingham

Ham, C. (1993a). How go the NHS reforms? *British Medical Journal*, **306**, 77–8.

Ham, C. (1993b). Reviewing the NHS Review. *British Medical Journal*, **307**, 5–6.

Ham, C. (1993c). *Locality Purchasing*. Birmingham: Health Services Management Centre, University of Birmingham

Howie, J.G.R. (1992). The Scottish general practice shadow fund-holding project – outline of an evaluation. *Health Bulletin*, **50**.

Howie, J.G.R. (1993). Evaluation of the Scottish shadow fund-holding project: first results. *Health Bulletin*, **51**, 2.

Howie, J.G.R. (1995). *Evaluation of the Scottish Shadow Fundholding Project: Final Report*. Edinburgh: Department of General Practice

Howland, J., Stokes, J., Crane, S.C. and Belanger, A.J. (1987). Adjusting capitation using chronic disease factors: a preliminary study. *Health Care Financing Review*, **9**, 2, 15–23.

Labour Party (1995). *Renewing the NHS*. London: Labour Party.

Lubitz, J., Beebe, J. and Riley, G. (1985). Improving the Medicare HMO payment formula to deal with biased selection. *Advances in Health Economics and Health Services Research* (edited by R. Scheffler and L. Rossiter), **6**, 101–22. Greenwich, Connecticut: JAI Press.

Luft, H.S. (1986). Compensating for biased selection in health insurance. *Milbank Quarterly*, **64**, 566–91.

Luft, H. and Miller, R.H. (1988). Patient selection in a competitive health system. *Health Affairs*, **7**, 3, 97–119.

Matsaganis, M. and Glennerster, H. (1994). The threat of cream skimming in the post-reform NHS. *Journal of Health Economics*, **13**, 31–60.

Maynard, A. (1986). Performance incentives in general practice. In: G. Teeling Smith (Ed.), *Health Education and General Practice*. London: Office of Health Economics.

Maxwell, M., Heaney, D., Howie, J.R.G. and Noble, S. (1993). General Practice Fundholding: observations on prescribing patterns and costs using the defined daily dose method. *British Medical Journal*, **307**, 1190–4.

National Audit Office (1994). *General Practitioner Fundholding in England*. London: HMSO.

Newhouse, J.P. (1982). Is competition the answer? *Journal of Health Economics*, **1**, 1, 109–16.

Newhouse, J.P. (1986). Rate adjusters for Medicare under capitation. *Health Care Financing Review*, Annual Supplement, 45–55.

Newhouse, J.P., Manning, W.G., Keeler, E.B. and Sloss, E.M. (1989). Adjusting capitation rates using objective health measures and prior utilization. *Health Care Financing Review*, **10**, 3, 41–54.

Propper, C. (1993). Quasi-markets contracts and quality in health and social care: the US experience. In: J. Le Grand and W. Bartlett (Eds), *Quasi-Markets and Social Policy*. London: Macmillan.

Robinson R. and Le Grand, J. (1994). *Evaluating the NHS Reforms*. London: Kings' Fund Institute.

Robinson, J.C., Luft, H.S., Gardner, L.B. and Morrison, E.M. (1991). A method for risk-adjusting employer contributions to competing health insurance plans. *Inquiry*, **28**, 107–16.

Sheldon, T., Smith, P., Borowitz, M., Martin, S. and Carr Hill, R. (1994). Attempt at deriving a formula for setting general practitioner fundholding budgets. *British Medical Journal*, **309**, 1059–64.

Stewart-Brown, S., Surrender, R., Bradlow, J., Coulter, A. and Doll, H. (1995). The effects of fundholding in general practice on prescribing habits three years after the introduction of the scheme. *British Medical Journal*, **311**, 1543–8.

Welch, W.P. (1985). Medicare capitation payments to HMOs in light of regression toward the mean in health care costs. *Advances in Health Economics and Health Services Research* (edited by R. Scheffler and L. Rossiter), **6**, 75–96. Greenwich, Connecticut: JAI Press.

Conclusions

INTRODUCTION

We have seen how very different health care systems have all tried to come to terms with setting limits to their budgets. In this concluding chapter we seek to identify some common themes and questions that have emerged. The first is perhaps the most fundamental.

SHOULD WE TRY?

In his contribution van de Ven (Chapter 5) asks why we should be seeking to set limits to health spending at all? We do not worry that individuals are spending "too much" on cars or video recorders or visits to the opera. His answer is that for these kinds of purchases it is up to the individual to decide his or her own budget limit in accordance with personal preferences. Health is different in that the state accepts some responsibility to ensure access to health care for its citizens and places a compulsory levy on them to finance such access. This requires some collective decision on what the scale of the compulsory levies shall be.

This is, of course, a rather Eurocentric view. The USA has yet to accept that responsibility for all its citizens. Nevertheless, the attempt to think how to do so in Clinton's programme of failed reforms sharpened American concern with ways of limiting the health budget. The Oregon experiment, moreover, began from just such a concern with the public budget for health care. Firms and other agencies that face similarly pressing calls on their budgets from their employees are also looking for ways of checking the rise in health costs, not only governments.

The moral, political and practical consequences of such an attempt have common features across all the varied systems our contributors discussed. The moral and political issues were elaborated, in particular, in Ham's contribution (Chapter 2). As he points out, rising health costs and the seemingly universal desire to reduce

Fixing Health Budgets: Experience from Europe and North America.
Edited by F. W. Schwartz, H.Glennerster and R. B. Saltman. Technical editor R. Busse.

individual tax burdens (or payment burdens on wages) will bring public health systems under increasing strain. Despite decades of movement towards universal and comprehensive coverage in such systems "something will have to give. This could be universality or comprehensiveness. It could be both."

WHAT BUDGETS?

In no country do governments attempt to fix the whole health care budget. No country tries to stop the ordinary person going into a pharmacist to buy aspirin, though they do put strict controls on what can be sold over the counter. None we studied made private health care illegal, though some countries did regulate it or put on quite rigorous price controls; Canada was one example, the Netherlands another.

Most insurance-based systems had found it particularly difficult to set budget limits. Caps had been set for particular types of expenditure such as hospital capital spending or drugs or the total spending of particular kinds of hospitals or for outpatient care. The German system is a particularly good example of this. Fixed budgets for the ambulatory care sector and for pharmaceuticals were introduced in Germany (Chapters 7 and 8). A less successful attempt with budgets of individual hospitals was set for hospital costs. Austria is even less well advanced. France began earlier but had considerable difficulty holding hospitals to their global budgets. The central European health systems, with their varied social partners, dispersed responsibilities and fee-for-service payment arrangements proved very difficult to contain. In countries that that had given universal access to health care for their populations and paid for these services out of taxation, political pressures forced action. This was most strikingly true of the UK with its centralized process of taxation and spending control. Smee describes this system in Chapter 11.

Yet restricting the state's spending in the UK has tended to increase the trend for individuals to insure themselves. At least this was so in the 1980s until the recession took effect. In France restrictions on the basic social security financed health budget lead to supplementary insurance funds being set up. We need to be clear whose health care we are restricting.

In the Netherlands, fixed budgets have included both public and private health expenditure because 33% of the Dutch population have private health insurance. As Maarse argues in Chapter 6: "The comprehensive approach rests upon the argument that private health expenditures have a similar effect upon employment, labour costs and competitiveness as public expenditures." To seek to contain the public system alone "would violate the principle of equity which is traditionally highly valued in Dutch health care."

FIXING FOR REAL?

Another key question is: How fixed is a fixed budget? For the purists the answer from most countries was – not very. Restrictions had held quite successfully, for example, in the Netherlands for a period, in Germany on some sectoral budgets for years, and in the UK and in Sweden for longer, but in all cases political pressures had forced some relaxation after several years. Yet no one believed attempts to fix budgets in some form would go away.

IMPACTS ON EFFICIENCY

The Canadian system has managed to combine both devolved responsibility for health finance and administration with a system of budget fixing that has been rather successful, as Nestman's chapter (Chapter 9) describes. But it also illustrates that this is only the beginning of the story. On their own, budget limits can remove institutions' incentives to act efficiently and adapt to consumer preferences. This was also a problem with the UK and Swedish models. As Maynard (Chapter 4) argues, for many years the UK Treasury was more interested in budget control than with the inefficiencies in the National Health Service. So long as it kept expenditure down, that was enough. Yet when the new reforms began to concentrate on achieving greater efficiency they were bought at a price – increasing growth in health costs. Getting the necessary improvements in the information needed for rationalizers to take informed choices has been a slow and costly process. The same road is being trodden much more gingerly by the Dutch but it is also proving very difficult politically.

PRIORITIES

If governments or private insurers try to set global budgets, who decides on the priorities within them? One approach is for the rationing to be undertaken centrally by the firm, state or province or nation state. A set of more or less essential treatments may be listed that will be covered by the insurance scheme or state health care system. There is a tension between leaving this to professionals or, instead, taking the political consequences of trying to reach a collective decision about such a hot issue. Even when brave politicians take it upon themselves to try to do so, there are difficulties in deciding what outcomes a society wishes to maximize – reduced mortality, morbidity, different kinds of ill health, the duration of a given condition or some combination of all of these. If a combination, what weights do you give to each of the above factors or others? In Chapter 3 Ganiats and Kaplan discuss these issues very clearly, the problems that beset Oregon in trying to grasp these issues and the sheer complexity and practical problems that have to be faced if one wishes to go down that road.

The easier way out is to fix global budgets for broad categories of institutional spending, set on the basis of past spending, and then let the professions make the rationing decisions or find, alternatively or additionally, ways of rationalizing their services. Most countries have taken this kind of route while edging incrementally towards shaving some treatments off the list that is publicly fundable.

WHOSE BUDGET?

This still leaves open the question: Whose budget are you setting? This has produced a lively debate in the UK where two answers have been produced and hence two competing solutions. These are, first, an authority representing the population of an area and, second, a family doctor purchasing services on behalf of his or her own patients. Both systems have their advantages. The first makes public health objectives rather easier to pursue and is cheaper to administer. The second puts the hospitals under much more pressure to produce effective and speedy results. Glennerster's contribution (Chapter 14) discusses this devolved system of budget setting in the UK.

A similar debate is taking place in Sweden and in the Netherlands. In the latter, a suggested reform would give patients the chance of belonging to competing insurers all receiving weighted capitation for each member who joined. Germany's "risk compensation scheme" tries to achieve the same effect. The sting is in the tail – what weights do we apply and are they sufficient to deter insurers from trying to reject the potentially expensive patient. HMOs in America, sickness funds in Germany and GP fundholders in the UK face a similar range of issues.

In the Netherlands, France and Germany the attempt is to apply limits to individual hospital spending while keeping a fee for service or fee for length of stay. The Netherlands appears to have been the most successful in this high wire act.

THE FUTURE

In the past decade health care systems throughout the world, but especially in Europe and North America, have made great strides in both conceptualizing the problems of cost containment but also in working through many of the problems. As the detailed contributions to this volume show, progress has been hard but real. The moral and technical issues overlap more than in most fields of economics and political science so it is not surprising that no one solution emerges. Each country's health care system is cherished for all its inefficiencies and inequities. Yet the fiscal pressures are building and will not let politicians, the medical profession or the public forget the issue.

Index

Index compiled by Liz Granger.